Plant Breeding

Plant Breeding

Kevin Sterling

MURPHY & MOORE
www.murphy-moorepublishing.com

Plant Breeding
Kevin Sterling
ISBN: 978-1-63987-431-6 (Hardback)

MURPHY & MOORE

Published by Murphy & Moore Publishing,
1 Rockefeller Plaza,
New York City, NY 10020, USA

Cataloging-in-Publication Data

Plant breeding / Kevin Sterling.
p. cm.
Includes bibliographical references and index.
ISBN 978-1-63987-431-6
1. Plant breeding. 2. Breeding. 3. Agriculture. I. Sterling, Kevin.
SB123 .P53 2022
631.52--dc23

For more information regarding Murphy & Moore Publishing and its products, please visit the publisher's website www.murphy-moorepublishing.com

TABLE OF CONTENTS

PREFACE

It is with great pleasure that I present this book. It has been carefully written after numerous discussions with my peers and other practitioners of the field. I would like to take this opportunity to thank my family and friends who have been extremely supporting at every step in my life.

Plant genes determine what traits the plant will have. The science of modifying these characteristic traits of plants to produce desired traits is known as plant breeding. It is done to improve the quality of nutrients present in plant products that are used by humans and animals. It is also used to enhance food security by making new varieties of crops that are disease resistant, regionally adapted, higher yielding, drought tolerant and adaptive to different environments. This field uses a variety of techniques. These range from simple methods like selection and deliberate interbreeding to complex molecular genetics techniques like reverse breeding and doubled haploids, genetic modification and marker assisted selection. This book elucidates the concepts and innovative models around prospective developments with respect to this discipline. Different approaches, evaluations, methodologies and studies on plant breeding have been included herein. This book, with its detailed analyzes and data, will prove immensely beneficial to professionals and students involved in this area at various levels.

The chapters below are organized to facilitate a comprehensive understanding of the subject:

Chapter - Introduction

The science of changing the traits of plants to produce preferred characteristics is known as plant breeding. These characteristics can be branched into two types - quantitative and qualitative characteristics. The topics elaborated in this chapter will help in gaining a better perspective of plant breeding.

Chapter - Methods and Technologies

Different methods and technologies are used in plant breeding such as doubled haploidy, genomic selection strategy, genome editing, plant tissue culture, cisgenesis, agroinfiltration, mutagenesis, etc. This chapter has been carefully written to provide an easy understanding of the methods and technologies used in plant breeding.

Chapter - Breeding for Pollinated Species

A range of breeding methods are used in relation to self-pollinated and cross-pollinated species such as mass selection, pure line selection, hybridization, etc. This chapter closely examines the subject of breeding for self and cross-pollinated species to provide an extensive understanding of the subject.

Chapter - Breeding for Stress Tolerance

Plant breeding is concerned with management of various stress tolerance such as drought stress tolerance, heat stress tolerance, waterlogging tolerance, cold stress tolerance, salinity tolerance, etc. This chapter discusses breeding for stress tolerance for a thorough understanding of the subject.

Chapter - Genetically Modified Plants

The plants whose DNA are modified through genetic engineering for developing new traits in crops are termed as genetically modified plants. It includes golden rice. This chapter sheds light on various genetically modified plants for an in-depth understanding of the subject.

Kevin Sterling

Introduction 1

- Plant Breeding
- Characteristics of Plant Breeding

The science of changing the traits of plants to produce preferred characteristics is known as plant breeding. These characteristics can be branched into two types – quantitative and qualitative characteristics. The topics elaborated in this chapter will help in gaining a better perspective of plant breeding.

Plant Breeding

Plant breeding is an application of genetic principles to produce plants that are more useful to humans. This is accomplished by selecting plants found to be economically or aesthetically desirable, first by controlling the mating of selected individuals, and then by selecting certain individuals among the progeny. Such processes, repeated over many generations, can change the hereditary makeup and value of a plant population far beyond the natural limits of previously existing populations.

'Double Delight,' a hybrid tea rose (Rosa 'Double Delight'). The flowers are highly fragrant and bloom continuously throughout the season.

Goals

The plant breeder usually has in mind an ideal plant that combines a maximum number of desirable characteristics. These characteristics may include resistance to diseases and insects; tolerance to heat, soil salinity, or frost; appropriate size, shape, and time to maturity; and many other general and specific traits that contribute to improved adaptation to the environment, ease in growing and handling, greater yield, and better quality. The breeder of horticultural plants must also consider aesthetic appeal. Thus the breeder can rarely focus attention on any one characteristic but must take into account the manifold traits that make the plant more useful in fulfilling the purpose for which it is grown. Plant breeding is an important tool in promoting global food security, and many staple crops have been bred to better withstand extreme weather conditions associated with global warming, such as drought or heat waves.

Increase of Yield

One of the aims of virtually every breeding project is to increase yield. This can often be brought about by selecting obvious morphological variants. One example is the selection of dwarf, early maturing varieties of rice. These dwarf varieties are sturdy and give a greater yield of grain. Furthermore, their early maturity frees the land quickly, often allowing an additional planting of rice or other crop the same year.

Another way of increasing yield is to develop varieties resistant to diseases and insects. In many cases the development of resistant varieties has been the only practical method of pest control. Perhaps the most important feature of resistant varieties is the stabilizing effect they have on production and hence on steady food supplies. Varieties tolerant to drought, heat, or cold provide the same benefit.

Modifications of Range and Constitution

Another common goal of plant breeding is to extend the area of production of a crop species. A good example is the modification of grain sorghum since its introduction to the United States in the 1750s. Of tropical origin, grain sorghum was largely confined to the southern Plains area and the Southwest, but earlier-maturing varieties were developed, and grain sorghum is now an important crop as far north as North Dakota.

Development of crop varieties suitable for mechanized agriculture has become a major goal of plant breeding in recent years. Uniformity of plant characters is very important in mechanized agriculture because field operations are much easier when the individuals of a variety are similar in time of germination, growth rate, size of fruit, and so on. Uniformity in maturity is, of course, essential when crops such as tomatoes and peas are harvested mechanically.

The nutritional quality of plants can be greatly improved by breeding. For example, it is possible to breed varieties of corn (maize) much higher in lysine than previously existing varieties. Breeding high-lysine maize varieties for those areas of the world where maize is the major source of this nutritionally essential amino acid has become a major goal in plant breeding. This "biofortification" of food crops, a term which also includes genetic modification, has been shown to improve nutrition and is especially useful in developing areas where nutritional deficiencies are common and medical infrastructure may be lacking.

In breeding ornamental plants, attention is paid to such factors as longer blooming periods, improved keeping qualities of flowers, general thriftiness, and other features that contribute to usefulness and aesthetic appeal. Novelty itself is often a virtue in ornamentals, and the spectacular, even the bizarre, is often sought.

Characteristics of Plant Breeding

Quantitative

Quantitative characters are governed by several genes. Hence they are also known as polygenic characters.

Such characters are expressed in degree rather than kind. Polygenic traits exhibit continuous variation from one extreme to other and therefore, grouping into distinct classes is not possible.

In such traits, effect of individual gene is small and undetectable. Polygenic traits are highly sensitive to environmental changes.

Such characters are measured in terms of height, length, width, weight, and duration. Statistical analysis of these characters is based on means, variances and co-variances.

Examples of quantitative characters are yield, days to flower, days to maturity, seed size, seed oil content, protein content etc.

Qualitative

Qualitative characters are controlled by one or few genes and therefore called monogenic or oligogenic traits.

Such characters exhibit discontinuous or discrete variation and therefore, can be easily grouped into distinct classes.

In such characters effect of individual gene is large and easily detectable. These characters are also known as major gene characters.

Oligogenic characters are very little influenced by environmental variation. Such characters are recorded in terms of shape, surface and color.

Statistical analysis of such traits is based on ratios and frequencies. Both quantitative and qualitative characters have some similarities, e.g. Segregations, mutation, linkage, reciprocal differences, dominance and epistasis etc. are common features of polygenic and oligogenic traits. In plant breeding, both oligogenic and polygenic characters are equally important.

2 Methods and Technologies

- Conventional Methods
- Biotechnological Methods
- Doubled Haploidy
- Molecular Breeding and Marker-assisted Selection
- CRISPR-Cas9
- Genomic Selection Strategy
- Genome Editing
- Plant Tissue Culture
- Cisgenesis
- Agroinfiltration
- Nuclease-based techniques
- Oligonucleotide-directed Mutagenesis-based Techniques
- Mutagenesis in Plant Breeding for Disease and Pest Resistance

Different methods and technologies are used in plant breeding such as doubled haploidy, genomic selection strategy, genome editing, plant tissue culture, cisgenesis, agroin-filtration, mutagenesis, etc. This chapter has been carefully written to provide an easy understanding of the methods and technologies used in plant breeding.

Conventional Methods

Conventional plant breeding methods resulting in hybrid varieties had a tremendous impact on agricultural productivity over the last decades. It develops new plant varieties by the process of selection and seeks to achieve expression of genetic material which is already present within the species.

Selection

Selection is the choice of certain individuals from a mixed population for one or more desirable traits. Selection is the oldest and basic method of plant breeding. There are two main types of Selection:

- Natural Selection: This is a rule in the nature and results in evolution reflected in the Darwinian principle "survival of the fittest". It takes longer time in bringing about desired variation.

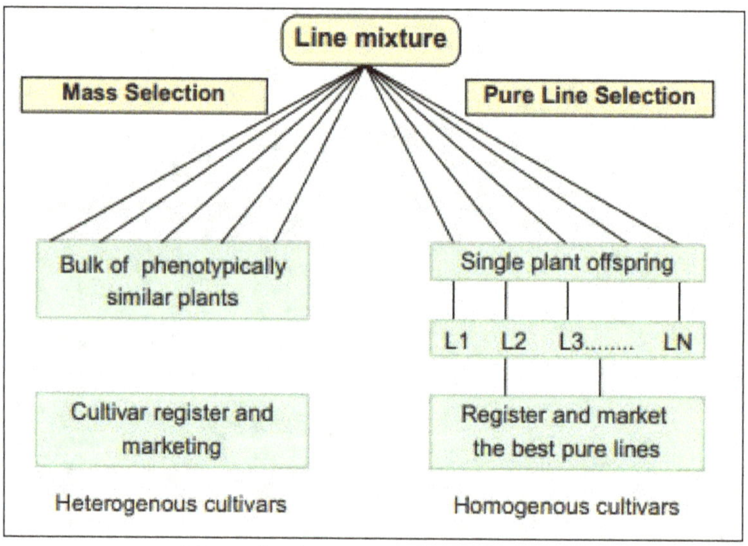

Mass selection vs. Pureline selection.

- Artificial Selection: It is a human involved process in having better crop from a mixed population where the individuals differ in character. The following are the three main types of artificial selection:
    - Mass Selection: In mass selection a large number of plants of similar phenotype or morphological characters are selected and their seeds are mixed together to constitute a new variety. The population obtained from the selected plants would be more uniform than the original population and are not individually tested. After repeated selection for about five to six years, selected seeds are multiplied and distributed to the farmers. The only disadvantage of mass selection is that it is difficult to distinguish the hereditary variation from environmental variation.
    - Pureline selection: Johannsen in 1903 coined the word pureline. It is a collection of plants obtained as a result of repeated self-pollination from a single homozygous individual. Hence, a variety formed by this method shows more homozygosity with respect to all genes. The disadvantage of this type is that the new genotypes are never created and they are less adaptable and less stable to the environmental fluctuations.

- Clonal Selection: In asexually propagated crop, progenies derived from a plant resemble in genetic constitution with the parent plant as they are mitotically divided. Based on their phenotypic appearance, clonal selection is employed to select improved variety from a mixed population (clones). The selected plants are multiplied through vegetative propagation to give rise to a clone. The genotype of a clone remains unchanged for a long period of time.

Hybridization

Hybridization is the method of producing new crop varieties in which two or more plants of unlike genetically constitution is crossed together that result in a progeny called hybrid. Hybridization offers improvement in crop and is the only effective means of combining together the desirable characters of two or more varieties or species. The first natural hybridization was observed by Cotton Mather in maize.

Steps in Hybridization

Steps involved in hybridization are as follows:

- Selection of Parents: Male and female plants of the desired characters are selected. It should be tested for their homozygosity.
- Emasculation: It is a process of removal of anthers to prevent self-pollination before anthesis (period of opening of a flower).
- Bagging: The stigma of the flower is protected against any undesirable pollen grains, by covering it with a bag.
- Crossing: Transfer of pollen grains from selected male flower to the stigma of the female emasculated flower.
- Harvesting seeds and raising plants: The pollination leads to fertilization and finally seed formation takes place. The seeds are grown into new generation which are called hybrid.

Types of Hybridization

According to the relationship between plants, the hybridization is divided into:

- Intravarietal hybridization: The cross between the plants of same variety. Such crosses are useful only in the self-pollinated crops.
- Intervarietal hybridization: The cross between the plants belonging to two different varieties of the same species and is also known as intraspecific hybridization. This technique has been the basis of improving self-pollinated as well as cross pollinated crops.

- Interspecific hybridization: The cross between the plants belonging to different species belonging to the same genus is also called intragenic hybridization. It is commonly used for transferring the genes of disease, insect, pest and drought resistance from one species to another. Example: Gossypium hirsutum × Gossypium arboreum – Deviraj.
- Intergeneric hybridization: The crosses are made between the plants belonging to two different genera. The disadvantages are hybrid sterility, time consuming and expensive procedure. Example: Raphanobrassica, Triticale.

Heterosis

Heterosis (hetero- different; sis - condition) G.H. Shull was the first scientist to use the term heterosis in 1912. The superiority of the F1 hybrid in performance over its parents is called heterosis or hybrid vigour. Vigour refers to increase in growth, yield, and greater adaptability of resistance to diseases, pest and drought. Vegetative propagation is the best suited measure for maintaining hybrid vigour, since the desired characters are not lost and can persist over a period of time. Many breeders believe that its magnitude of heterosis is directly related to the degree of genetic diversity between the two parents. Depending on the nature, origin, adaptability and reproducing ability heterosis can be classified as:

- Euheterosis: This is the true heterosis which is inherited and is further classified as:
    - Mutational Euheterosis: Simplest type of euheterosis and results from the sheltering or eliminating of the deleterious, unfavourable often lethal, recessive, mutant genes by their adaptively superior dominant alleles in cross pollinated crops.
    - BalancedEuheterosis: Well-balanced gene combinations which is more adaptive to environmental conditions and agricultural usefulness.
- Psuedoheterosis: Also termed as luxuriance. Progeny possess superiority over parents in vegetative growth but not in yield and adaptation, usually sterile or poorly fertile.

Mutation Breeding

Muller and Stadler coined the term mutation breeding. It represents a new method of conventional breeding procedures as they have the advantage of improving the defect without losing agronomic and quality character in agriculture and crop improvement. Mutation means the sudden heritable changes in the genotype or phenotype of an organism. Gene mutations are of considerable importance in plant breeding as they provide essential inputs for evolution as well as for re-combination and selection. It is the only method for improving seedless crops.

Radiation such as UV short wave, X-ray, Alpha (α), Beta (β), Gamma waves and many chemicals such as cesium, EMS (ethyl methane sulfonate), nitromethyl, urea induces mutation to develop new variety of crops. Example: Triple gene dwarf wheat with increase in yield and height. Atomita 2 - rice with saline tolerance and pest resistance.

Polyploid Breeding

Majority of flowering plants are diploid (2n). The plants which possess more than two sets of chromosome are called polyploids. Polyploidy is a major force in the evolution of both wild and cultivated plants. Polyploidy often exhibit increased hybrid vigour increased heterozygosity, increase the tolerance to both biotic and abiotic stresses, buffering of deleterious mutations. In addition, polyploidy often results in reduced fertility due to meiotic error allowing the production of seedless varieties.

When chromosome number is doubled by itself in the same plant, is called autopolyploidy. Example: A triploid condition in sugarbeets, apples and pear has resulted in the increase in vigour and fruit size, large root size, large leaves, flower, more seeds and sugar content in them. It also resulted in seedless tomato, apple, watermelon and orange. Polyploidy can be induced by the use of colchicine to double the chromosome number. Allopolyploids are produced by multiplication of chromosome sets that are initially derived from two different species. Example: Triticale (Triticum durum x secale cereale) Raphanobrassica (Brassica oleraceae x Raphanus sativus).

Biotcchnological Methods

Biotechnology is the discipline which deals with the use of living organisms or their products. In this wide sense, also traditional agriculture may be seen as a form of biotechnology. The European Federation of Biotechnology defines biotechnology as "the integrated use of biochemistry, microbiology and engineering sciences in order to achieve technological (industrial) application of the capability of microorganisms, cultured tissue cells and parts thereof". In recent years, biotechnology has developed rapidly as a practical means for accelerating success in plant breeding and improving economically important crops.

In Vitro Cultivation

In Vitro Cultivation of Plant Cells and Regeneration of Plants from Cultured Cells Certain isolated somatic plant cells can be cultured in vitro (in the test tube) and are capable of proliferation and organization into tissues and eventually into complete plants. The process of regenerating whole plants out of plant cells is called in vitro regeneration. The three factors affecting plant regeneration are genotype, explant source, and culture conditions, including culture medium and environment. Different mixtures of plant hormones and other compounds in varying concentrations are used to achieve regeneration of plants from cultured cells and tissues. As the plant hormonal mechanisms are not yet understood completely, the development of in vitro cultivation and regeneration systems is still largely based on empirically testing variations of the three above mentioned factors.

In Vitro Selection and Somaclonal Variation

Plants regenerated from cell cultures may exhibit phenotypes differing from their parent plants, sometimes at quite high frequencies. If these are heritable and affecting desirable agronomic traits, such "somaclonal variation" can be incorporated into breeding programs. However, finding of specific valuable traits by this method is largely left to chance and hence inefficient. Rather than relying on this undirected process, selection in vitro targets specific traits by subjecting large populations of cultured cells to the action of a selective agent in the Petri dish. For purposes of disease resistance, this selection can be provided by pathogens, or isolated pathotoxins that are known to have a role in pathogenesis. The selection will only allow those cells to survive and proliferate that are resistant to the challenge. Selection of cells also plays an important role in genetic engineering, where special marker genes are used to select for transgenic cells.

Somatic Hybrid Plants

Somatic hybrid plants are plants derived from the fusion of somatic cells. Cell fusion was developed after the successful culture of a large number of plant cells stripped of their cell walls. The resulting cells without walls are referred to as protoplasts. Since also protoplasts from phylogenetically unrelated species can be fused, attempts have been made to overcome sexual incompatibility using protoplast fusion. In most cases, these attempts failed because growth and division of the fused cells did not take place when only distantly related cells were fused. Successful fusions between sexually incompatible petunia species and between potatoes and tomatoes did not lead to economically interesting products, but important contributions to the understanding of cell wall regeneration and other mechanisms were achieved.

Doubled Haploidy

A doubled haploid (DH) is a genotype formed when haploid cells undergo chromosome doubling. Artificial production of doubled haploids is important in plant breeding.

Haploid cells are produced from pollen or egg cells or from other cells of the gametophyte, then by induced or spontaneous chromosome doubling, a doubled haploid cell is produced, which can be grown into a doubled haploid plant. If the original plant was diploid, the haploid cells are monoploid, and the term doubled monoploid may be used for the doubled haploids. Haploid organisms derived from tetraploids or hexaploids are sometimes called dihaploids (and the doubled dihaploids are, respectively, tetraploid or hexaploid).

Conventional inbreeding procedures take six generations to achieve approximately complete homozygosity, whereas doubled haploidy achieves it in one generation. Dihaploid plants derived from tetraploid crop plants may be important for breeding programs that involve diploid wild relatives of the crops.

Production of Doubled Haploids

Doubled haploids can be produced in vivo or in vitro. Haploid embryos are produced in vivo by parthenogenesis, pseudogamy, or chromosome elimination after wide crossing. The haploid embryo is rescued, cultured, and chromosome-doubling produces doubled haploids. The in vitro methods include gynogenesis (ovary and flower culture) and androgenesis (anther and microspore culture). Androgenesis is the preferred method. Another method of producing the haploids is wide crossing. In barley, haploids can be produced by wide crossing with the related species Hordeum bulbosum; fertilization is affected, but during the early stages of seed development the H. bulbosum chromosomes are eliminated leaving a haploid embryo. In tobacco (Nicotiana tabacum), wide crossing with Nicotiana africana is widely used. When N. africana is used to pollinate N. tabacum, 0.25 to 1.42 percent of the progeny survive and can readily be identified as either F1 hybrids or maternal haploids. Although these percentages appear small, the vast yield of tiny seeds and the early death of most seedlings provide significant numbers of viable hybrids and haploids in relatively small soil containers. This method of interspecific pollination serves as a practical way of producing seed-derived haploids of N. tabacum, either as an alternative method or complementary method to anther culture.

Genetics of DH Population

In DH method only two types of genotypes occur for a pair of alleles, A and a, with the frequency of ½ AA and ½ aa, while in diploid method three genotypes occur with the frequency of ¼ AA, ½ Aa, ¼ aa. Thus, if AA is desirable genotype, the probability of obtaining this genotype is higher in haploid method than in diploid method. If n loci are segregating, the probability of getting the desirable genotype is (1/2)n by the haploid method and (1/4)n by the diploid method. Hence the efficiency of the haploid method is high when the number of genes concerned is large.

Studies were conducted comparing DH method and other conventional breeding methods and it was concluded that adoption of doubled haploidy does not lead to any bias of genotypes in populations, and random DHs were even found to be compatible to selected line produced by conventional pedigree method.

Applications of DHs Plant Breeding

Mapping Quantitative Trait Loci

Most of the economic traits are controlled by genes with small but cumulative effects. Although the potential of DH populations in quantitative genetics has been understood for some time, it was the advent of molecular marker maps that provided the impetus for their use in identifying loci controlling quantitative traits. As the quantitative trait loci (QTL) effects are small and highly influenced by environmental factors, accurate phenotyping with replicated trials is needed. This is possible with doubled haploidy organisms because of their true breeding nature and because they can conveniently be produced in large numbers. Using DH populations, 130 quantitative traits have been mapped in nine crop species. In total, 56 DH populations were used for QTL detection.

Backcross Breeding

In backcross conversion, genes are introgressed from a donor cultivar or related species into a recipient elite line through repeated backcrossing. A problem in this procedure is being able to identify the lines carrying the trait of interest at each generation. The problem is particularly acute if the trait of interest is recessive, as it will be present only in a heterozygous condition after each backcross. The development of molecular markers provides an easier method of selection based on the genotype (marker) rather than the phenotype. Combined with doubled haploidy it becomes more effective. In marker assisted backcross conversion, a recipient parent is crossed with a donor line and the hybrid (F1) backcrossed to the recipient. The resulting generation (BC1) is backcrossed and the process repeated until the desired genotypes are produced. The combination of doubled haploidy and molecular marker provides the short cut. In the back cross generation one itself a genotype with the character of interest can be selected and converted into homozygous doubled haploid genotype. Chen used marker assisted backcross conversion with doubled haploidy of BC1 individuals to select stripe rust resistant lines in barley.

Bulked Segregant Analysis

In bulked segregant analysis, a population is screened for a trait of interest and the genotypes at the two extreme ends form two bulks. Then the two bulks are tested for the presence or absence of molecular markers. Since the bulks are supposed to contrast in the alleles that contribute positive and negative effects, any marker polymorphism between the two bulks indicates the linkage between the marker and trait of interest. BSA is dependent on accurate phenotyping and the DH population has particular advantage in that they are true breeding and can be tested repeatedly. DH populations are commonly used in bulked segregant analysis, which is a popular method in marker assisted breeding. This method has been applied mostly to rapeseed and barley.

Genetic Maps

Genetic maps are very important to understand the structure and organization of genomes from which evolution patterns and syntenic relationships between species can be deduced. Genetic maps also provide a framework for the mapping of genes of interest and estimating the magnitude of their effects and aid our understanding of genotype/phenotype associations. DH populations have become standard resources in genetic mapping for species in which DHs are readily available. Doubled haploid populations are ideal for genetic mapping. It is possible to produce a genetic map within two years of the initial cross regardless of the species. Map construction is relatively easy using a DH population derived from a hybrid of two homozygous parents as the expected segregation ratio is simple, i.e. 1:1. DH populations have now been used to produce genetic maps of barley, rapeseed, rice, wheat, and pepper. DH populations played a major role in facilitating the generation of the molecular marker maps in eight crop species.

Genetic Studies

Genetic ratios and mutation rates can be read directly from haploid populations. A small doubled haploid (DH) population was used to demonstrate that a dwarfing gene in barley is located chromosome 5H. In another study the segregation of a range of markers has been analyzed in barley.

Genomics

Although QTL analysis has generated a vast amount of information on gene locations and the magnitude of effects on many traits, the identification of the genes involved has remained elusive. This is due to poor resolution of QTL analysis. The solution for this problem would be production of recombinant chromosome substitution line, or stepped aligned recombinant inbred lines. Here, backcrossing is carried out until a desired level of recombination has occurred and genetic markers are used to detect desired recombinant chromosome substitution lines in the target region, which can be fixed by doubled haploidy. In rice, molecular markers have been found to be linked with major genes and QTLs for resistance to rice blast, bacterial blight, and sheath blight in a map produced from DH population.

Elite Crossing

Traditional breeding methods are slow and take 10–15 years for cultivar development. Another disadvantage is inefficiency of selection in early generations because of heterozygosity. These two disadvantages can be over come by DHs, and more elite crosses can be evaluated and selected within less time.

Cultivar Development

Uniformity is a general requirement of cultivated line in most species, which can be easily obtained through DH production. There are various ways in which DHs can be used in cultivar production. The DH lines themselves can be released as cultivars, they may be used as parents in hybrid cultivar production or more indirectly in the creation of breeders lines and in germplasm conservation. Barley has over 100 direct DH cultivars. According to published information there are currently around 300 DH derived cultivars in 12 species worldwide.

The relevance of DHs to plant breeding has increased markedly in recent years owing to the development of protocols for 25 species. Doubled haploidy already plays an important role in hybrid cultivar production of vegetables, and the potential for ornamental production is being vigorously examined. DHs are also being developed in the medicinal herb Valeriana officinalis to select lines with high pharmacological activity. Another interesting development is that fertile homozygous DH lines can be produced in species that have self-incompatibility systems.

Advantages of DHs

The ability to produce homozygous lines after a single round recombination saves a lot of time for the plant breeders. Studies conclude that random DH's are comparable to the selected lines in pedigree inbreeding. The other advantages include development of large number of homozygous lines, efficient genetic analysis and development of markers for useful traits in much less time. More specific benefits include the possibility of seed propagation as an alternative to vegetative multiplication in ornamentals, and in species such as trees in which long life cycles and inbreeding depression preclude traditional breeding methods, doubled haploidy provides new alternatives.

Disadvantages of DHs

The main disadvantage with the DH population is that selection cannot be imposed on the population. But in conventional breeding selection can be practised for several generations: thereby desirable characters can be improved in the population.

In haploids produced from anther culture, it is observed that some plants are aneuploids and some are mixed haploid-diploid types. Another disadvantage associated with the double haploidy is the cost involved in establishing tissue culture and growth facilities. The over-usage of doubled haploidy may reduce genetic variation in breeding germplasm. Hence one has to take several factors into consideration before deploying doubled haploidy in breeding programmes.

Technological advances have now provided DH protocols for most plant genera. The number of species amenable to doubled haploidy has reached a staggering 250 in just a few decades. Response efficiency has also improved with gradual removal of species from recalcitrant category. Hence it will provide greater efficiency of plant breeding.

Molecular Breeding and Marker-assisted Selection

Marker assisted selection or marker aided selection (MAS) is an indirect selection process where a trait of interest is selected based on a marker (morphological, biochemical or DNA/RNA variation) linked to a trait of interest (e.g. productivity, disease resistance, abiotic stress tolerance, and quality), rather than on the trait itself. This process has been extensively researched and proposed for plant and animal breeding, nevertheless, as of 2013 "breeding programs based on DNA markers for improving quantitative traits in plants are rare".

For example, using MAS to select individuals with disease resistance involves identifying a marker allele that is linked with disease resistance rather than the level of disease resistance. The assumption is that the marker associates at high frequency with the gene or quantitative trait locus (QTL) of interest, due to genetic linkage (close proximity, on the chromosome, of the marker locus and the disease resistance-determining locus). MAS can be useful to select for traits that are difficult or expensive to measure, exhibit low heritability and are expressed late in development. At certain points in the breeding process the specimens are examined to ensure that they express the desired trait.

Positive and Negative Selectable Markers

The following terms are generally less relevant to discussions of MAS in plant and animal breeding, but are highly relevant in molecular biology research:

- Positive selectable markers are selectable markers that confer selective advantage to the host organism. An example would be antibiotic resistance, which allows the host organism to survive antibiotic selection.
- Negative selectable markers are selectable markers that eliminate or inhibit growth of the host organism upon selection. An example would be thymidine kinase, which makes the host sensitive to ganciclovir selection.

A distinction can be made between selectable markers (which eliminate certain genotypes from the population) and screenable markers (which cause certain genotypes to be readily identifiable, at which point the experimenter must "score" or evaluate the population and act to retain the preferred genotypes). Most MAS uses screenable markers rather than selectable markers.

Gene vs. Marker

The gene of interest directly causes production of protein s or RNA that produce a desired trait or phenotype, whereas markers (a DNA sequence or the morphological or biochemical markers produced due to that DNA) are genetically linked to the gene of interest. The gene of interest and the marker tend to move together during segregation of gametes due to their proximity on the same chromosome and concomitant reduction in recombination (chromosome crossover events) between the marker and gene of interest. For some traits, the gene of interest has been discovered and the presence of desirable alleles can be directly assayed with a high level of confidence. However, if the gene of interest is not known, markers linked to the gene of interest can still be used to select for individuals with desirable alleles of the gene of interest. When markers are used there may be some inaccurate results due to inaccurate tests for the marker. There also can be false positive results when markers are used, due to recombination between the marker of interest and gene (or QTL). A perfect marker would elicit no false positive results. The term 'perfect marker' is sometimes used when tests are performed to detect a SNP or other DNA polymorphism in the gene of interest, if that SNP or other polymorphism is the direct cause of the trait of interest. The term 'marker' is still appropriate to use when directly assaying the gene of interest, because the test of genotype is an indirect test of the trait or phenotype of interest.

Important Properties of Ideal Markers for MAS

An ideal marker:

- Easy recognition of all possible phenotypes (homo- and heterozygotes) from all different alleles.
- Demonstrates measurable differences in expression between trait types or gene of interest alleles, early in the development of the organism.
- Testing for the marker does not have variable success depending on the allele at the marker locus or the allele at the target locus (the gene of interest that determines the trait of interest).
- Low or null interaction among the markers allowing the use of many at the same time in a segregating population.
- Abundant in number.
- Polymorphic.

Drawbacks of Morphological Markers

Morphological markers are associated with several general deficits that reduce their usefulness including:

- The delay of marker expression until late into the development of the organism.
- Dominance.

- Deleterious effects.
- Pleiotropy.
- Confounding effects of genes unrelated to the gene or trait of interest but which also affect the morphological marker (epistasis).
- Rare polymorphism.
- Frequent confounding effects of environmental factors which affect the morphological characteristics of the organism.

To avoid problems specific to morphological markers, DNA-based markers have been developed. They are highly polymorphic, exhibit simple inheritance (often codominant), are abundant throughout the genome, are easy and fast to detect, exhibit minimum pleiotropic effects, and detection is not dependent on the developmental stage of the organism. Numerous markers have been mapped to different chromosomes in several crops including rice, wheat, maize, soybean and several others, and in livestock such as cattle, pigs and chickens. Those markers have been used in diversity analysis, parentage detection, DNA fingerprinting, and prediction of hybrid performance. Molecular markers are useful in indirect selection processes, enabling manual selection of individuals for further propagation.

Selection for Major Genes Linked to Markers

'Major genes' that are responsible for economically important characteristics are frequent in the plant kingdom. Such characteristics include disease resistance, male sterility, self-incompatibility, and others related to shape, color, and architecture of whole plants and are often of mono- or oligogenic in nature. The marker loci that are tightly linked to major genes can be used for selection and are sometimes more efficient than direct selection for the target gene. Such advantages in efficiency may be due for example, to higher expression of the marker mRNA in such cases that the marker is itself a gene. Alternatively, in such cases that the target gene of interest differs between two alleles by a difficult-to-detect single nucleotide polymorphism, an external marker (be it another gene or a polymorphism that is easier to detect, such as a short tandem repeat) may present as the most realistic option.

Situations that are Favorable for Molecular Marker Selection

There are several indications for the use of molecular markers in the selection of a genetic trait. In such situations that:

- The selected character is expressed late in plant development, like fruit and flower features or adult characters with a juvenile period (so that it is not necessary to wait for the organism to become fully developed before arrangements can be made for propagation).

- The expression of the target gene is recessive (so that individuals which are heterozygous positive for the recessive allele can be crossed to produce some homozygous offspring with the desired trait).
- There is requirement for the presence of special conditions in order to invoke expression of the target genes, as in the case of breeding for disease and pest resistance (where inoculation with the disease or subjection to pests would otherwise be required). this advantage derives from the errors due to unreliable inoculation methods and the fact that field inoculation with the pathogen is not allowed in many areas for safety reasons. moreover, problems in the recognition of the environmentally unstable genes can be eluded.
- The phenotype is affected by two or more unlinked genes (epistatis). for example, selection for multiple genes which provide resistance against diseases or insect pests for gene pyramiding.

The cost of genotyping (an example of a molecular marker assay) is reducing while the cost of phenotyping is increasing particularly in developed countries thus increasing the attractiveness of MAS as the development of the technology continues.

Steps for MAS

Generally the first step is to map the gene or quantitative trait locus (QTL) of interest first by using different techniques and then using this information for marker assisted selection. Generally, the markers to be used should be close to gene of interest (<5 recombination unit or cM) in order to ensure that only minor fraction of the selected individuals will be recombinants. Generally, not only a single marker but rather two markers are used in order to reduce the chances of an error due to homologous recombination. For example, if two flanking markers are used at same time with an interval between them of approximately 20cM, there is higher probability (99%) for recovery of the target gene.

Use of MAS for Backcross Breeding

A minimum of five or six-backcross generations are required to transfer a gene of interest from a donor (may not be adapted) to a recipient (recurrent – adapted cultivar). The recovery of the recurrent genotype can be accelerated with the use of molecular markers. If the F1 is heterozygous for the marker locus, individuals with the recurrent parent allele s at the marker locus in first or subsequent backcross generations will also carry a chromosome tagged by the marker.

Marker Assisted Gene Pyramiding

Gene pyramiding has been proposed and applied to enhance resistance to disease and insects by selecting for two or more than two genes at a time. For example, in rice such pyramids have been developed against bacterial blight and blast. The advantage of use of markers in this case allows to select for QTL-allele-linked markers that have same phenotypic effect.

DNA Markers

Those characters which can be easily identified are called marker characters. Any genetic element (locus, allele, DNA sequence or chromosome feature) which can be readily detected by phenotype, cytological or molecular techniques, and used to follow a chromosome or chromosomal segment during genetic analysis is referred to as marker.

Markers related to variations in DNA fragments generated by restriction endonuclease enzymes are called DNA markers or genetic markers.

- Any unique DNA sequence which can be used in DNA hybridization, PCR or restriction mapping experiments to identify that sequence is called DNA marker,
- A gene or DNA sequence having a known location on a chromosome and associated with a particular gene or trait refers to DNA marker.

Markers are of four types:

- Morphological,
- Biochemical,
- Cytological,
- DNA markers.

Morphological

In plant breeding, markers that are related to variation in shape, size, colour and surface of various plant parts are called morphological markers. Such markers refer to available gene loci that have obvious impact on morphology of plant. Genes that affect form, coloration, male sterility or resistance among others have been analyzed in many plant species.

In rice, examples of this type of marker may include the presence or absence of awn, leaf sheath coloration, height, grain color, aroma etc. In well-characterized crops like maize, tomato, pea, barley or wheat, tens or even hundreds of such genes have been assigned to different chromosomes.

There are several demerits of morphological markers as given below:

- They generally express late into the development of an organism. Hence their detection is dependent on the development stage of the organism.
- They usually exhibit dominance.
- Sometimes they exhibit deleterious effects.

- They exhibit pleiotropy.
- They exhibit epistasis.
- They exhibit less polymorphism.
- They are highly influenced by the environmental factors.

Biochemical

Markers that are related to variation in proteins and amino acid banding pattern are known as biochemical markers. A gene encodes a protein that can be extracted and observed; for example, isozymes and storage proteins.

Cytological

Markers that are related to variation in chromosome number, shape, size and banding pattern are referred to as cytological markers. In other words, it refers to the chromosomal banding produced by different stains; for example, G banding.

A gene or other fragment of DNA whose location in the genome is known is called DNA marker. It is a unique (DNA sequence), occurring in proximity to the gene or locus of interest. It refers to any unique DNA sequence which can be used in DNA hybridization, PCR or restriction mapping experiments to identify that sequence.

It can be identified by a range of molecular techniques such as RFLPs, RAPDs, AFLP, SNPs, SCARs, microsatellites etc.

DNA markers are also known as molecular markers or genetic markers. To overcome problems associated with morphological markers, the DNA-based markers have been developed. Advantages of DNA markers are presented below:

- They are highly polymorphic.
- They have simple inheritance (often co-dominant).
- They abundantly occur throughout the genome.
- They are easy and fast to detect.
- They exhibit minimum pleiotropic effect.

Their detection is not dependent on the developmental stage of the organism.

Table: Comparison of Morphological Markers and DNA Markers.

S.no	Particulars	Morphological Markers	DNA Markers
1	Nature	Dominant	Co-dominant

2	Polymorphism	Low	High
3	Occurrence	Low	Abundant
4	Detection	Easy	Easy
5	Pleiotropic effect	High	Minimum
6	Epistasis	Present	Absent
7	Environmental effect	High	No effect

Properties of DNA Marker

An ideal DNA marker should have some properties or characteristics.

Polymorphism

Markers should exhibit high level of polymorphism. In other words, there should be variability in the markers. It should demonstrate measurable differences in expression between trait types and gene of interest.

Co-dominant

Marker should be co-dominant. It means, there should be absence of intra-locus interaction. It helps in identification of heterozygotes from homozygotes.

Multi-allelic

The marker should be multi-allelic. It useful in getting more variability/polymorphism for a character.

No Epistasis

There should be absence of epistasis. It makes Identification of all phenotypes (homo- and heterozygotes) easy.

Neutral

The marker should be neutral. The substitution of alleles at the marker locus should not alter the phenotype of an individual. This property is found in almost all the DNA markers.

No Effect of Environment

Markers should be insensitive to environment. This property is also found in almost all the DNA markers.

Applications of DNA Marker in Crop Improvement

DNA markers have several useful applications in crop improvement. The important applications are listed as follows:

- DNA markers are useful in the assessment of genetic diversity in germplasm, cultivars and advanced breeding material.
- DNA markers can be used for constructing genetic linkage maps.
- DNA markers are useful in identification of new useful alleles in the germplasm and wild species of crop plants.
- DNA markers are used in the marker assisted or marker aided selection. MAS has several advantages over straight selection.
- DNA markers are useful in the study of crop evolution.

RAPD

RAPD stands for 'Random Amplification of Polymorphic DNA'. It is a type of PCR, but the segments of DNA that are amplified are random. The scientist performing RAPD creates several arbitrary, short primers (8–12 nucleotides), then proceeds with the PCR using a large template of genomic DNA, hoping that fragments will amplify. By resolving the resulting patterns, a semi-unique profile can be gleaned from an RAPD reaction.

No knowledge of the DNA sequence of the targeted genome is required, as the primers will bind somewhere in the sequence, but it is not certain exactly where. This makes the method popular for comparing the DNA of biological systems that have not had the attention of the scientific community, or in a system in which relatively few DNA sequences are compared (it is not suitable for forming a cDNA databank). Because it relies on a large, intact DNA template sequence, it has some limitations in the use of degraded DNA samples. Its resolving power is much lower than targeted, species-specific DNA comparison methods, such as short tandem repeats. In recent years, RAPD has been used to characterize, and trace, the phylogeny of diverse plant and animal species.

RAPD markers are decamer (10 nucleotides long) DNA fragments from PCR amplification of random segments of genomic DNA with single primer of arbitrary nucleotide sequence and which are able to differentiate between genetically distinct individuals, although not necessarily in a reproducible way. It is used to analyse the genetic diversity of an individual by using random primers. Due to problems in experiment reproducibility, many scientific journals do not accept experiments merely based on RAPDs anymore. RAPD requires only one primer for amplification.

Working of RAPD

Unlike traditional PCR analysis, RAPD does not require any specific knowledge of the DNA sequence of the target organism: the identical 10-mer primers will or will not amplify a segment of DNA, depending on positions that are complementary to the primers' sequence. For example, no fragment is produced if primers annealed too far apart or 3' ends of the primers are not facing each other. Therefore, if a mutation has occurred in the template DNA at the site that was previously complementary to the primer, a PCR product will not be produced, resulting in a different pattern of amplified DNA segments on the gel.

Limitations

- Nearly all RAPD markers are dominant, i.e. it is not possible to distinguish whether a DNA segment is amplified from a locus that is heterozygous (1 copy) or homozygous (2 copies). Codominant RAPD markers, observed as different-sized DNA segments amplified from the same locus, are detected only rarely.
- PCR is an enzymatic reaction, therefore, the quality and concentration of template DNA, concentrations of PCR components, and the PCR cycling conditions may greatly influence the outcome. Thus, the RAPD technique is notoriously laboratory dependent and needs carefully developed laboratory protocols to be reproducible.
- Mismatches between the primer and the template may result in the total absence of PCR product as well as in a merely decreased amount of the product. Thus, the RAPD results can be difficult to interpret unlike traditional PCR analysis.

Applications of RAPD Marker in Plants

RAPD is used to distinguish between variety is based on difference in DNA sequence. RAPD have been used to identify nearly 15 commercial sunflower varieties. The new bean varieties (Phaseolus vulgaris) which are difficult to distinguish based on morphological trait have been used as ideal candidate for the application of RAPD marker methods when DNA was extracted from each varieties. The 12 samples were analysed using 60 primers.

This produced 296 markers and that could be scored. Almost 85% similarity was predicted. For example, DNA from plant allows the amplification of sequences a, c, d but not b. This indicates that in plant 1, primer sites for the primers used not found at sequence b. Similarly, a DNA sequence alternation at one of the primer binding site (priming) for a sequence 'a' has prevent it from being amplified when DNA from plant 2 is used.

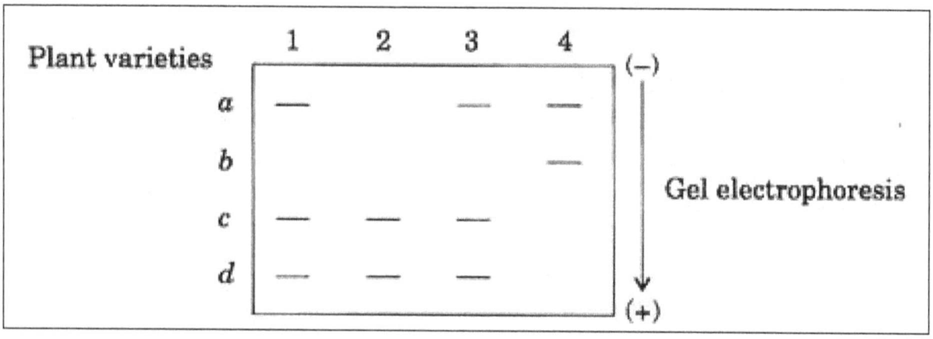

RAPD have been extensively used for number of horticultural crops in variety identification, genetic purity and sex determination. A specific RAPD marker has been used to select for high and low β-glucan content between barley varieties.

RAPD markers are employed in the construction of genetic maps. Genetic maps of several plants including model plant Arabidopsis and tobacco have been constructed. RAPD markers have been used to construct 15 linkage groups in coffee. Both genomic and chloroplast DNA provided the source of probes.

RAPD markers are used for the selection of segregating populations more or less indirectly, during plant breeding. These markers also accelerate back crossing process and allow the selection of individually with more of recurrent genome at each generation facilitate breeding programme to be completed within few generations.

RAPD molecular marker used in the direct selection of desirable trait. Molecular marker linked to the trait of interest can be screened for at any stage in the breeding programme.

RAPD and other molecular markers have great value in the selection for desirable trait in long-lived species which takes long time for maturity and show phenotypic character. For example, avacado (Persea americana) fruit quality can be assessed in seedling itself using RAPD molecular marker.

RAPD markers have been used to identify several disease resistant genes in plants. The rp94 gene is responsible for resistance to stem rust (Puccinia gramnis) in barley. RAPD markers identified to link to this gene. Similarly, RAPD markers linked to heat smut resistance gene have been characterized. Controlling of height in barley plant by specific gene has been used to locate dwarfism gene by RAPD marker.

In tissue culture work, somatic hybrids involving protoplast fusion requires thorough screening. However, screening of somatic hybrid is cumbersome. Therefore, RAPD markers can be exploited in identifying somatic hybrids. RAPD analysis provides an important tool for the characterization of biodiversity.

Identification of areas rich in endemic genotypes helps in habitat conservative and prevents species extinction. Molecular analysis of genetic diversity using RAPD or RFLP in plant genetic germplasm collection facilitate better management especially space and resources are serious constraints.

RAPD analysis has been used for the identification of duplicates in germplasm. These duplicates are then discarded once no morphological differences were detected. RAPD analysis has been implicated in the analysis of rice genome collections held at the International Rice Research Institute, Phillipines.

RAPD have also been used in variety identification and purity in grain processing in food industry. For example, particular duran variety of wheat is used in the preparation of food products. Contamination of other variety can be identified by using these molecular markers.

Demerits of RAPD

- Constraints about reproducibility of results.
- Since RAPD markers are dominant, only half the genetic informations are co-dominant markers.
- Null alleles not directly detected.

RFLP

Restriction fragment length polymorphism (RFLP) is a technique that exploits variations in homologous DNA sequences, known as polymorphisms, in order to distinguish individuals, populations, or species or to pinpoint the locations of genes within a sequence. The term may refer to a polymorphism itself, as detected through the differing locations of restriction enzyme sites, or to a related laboratory technique by which such differences can be illustrated. In RFLP analysis, a DNA sample is digested into fragments by one or more restriction enzymes, and the resulting restriction fragments are then separated by gel electrophoresis according to their size.

Although now largely obsolete due to the emergence of inexpensive DNA sequencing technologies, RFLP analysis was the first DNA profiling technique inexpensive enough to see widespread application. RFLP analysis was an important early tool in genome mapping, localization of genes for genetic disorders, determination of risk for disease, and paternity testing.

RFLP Analysis

The basic technique for the detection of RFLPs involves fragmenting a sample of DNA with the application of a restriction enzyme, which can selectively cleave a DNA molecule wherever a short, specific sequence is recognized in a process known as a restriction digest. The DNA fragments produced by the digest are then separated by length through a process known as agarose gel electrophoresis and transferred to a membrane via the Southern blot procedure. Hybridization of the membrane to a labeled DNA probe then determines the length of the fragments which are complementary to the probe. A restriction fragment length polymorphism is said to occur when the length of a detected fragment varies between individuals, indicating non-identical sequence homologies. Each fragment length is considered an allele, whether it actually contains a coding region or not, and can be used in subsequent genetic analysis.

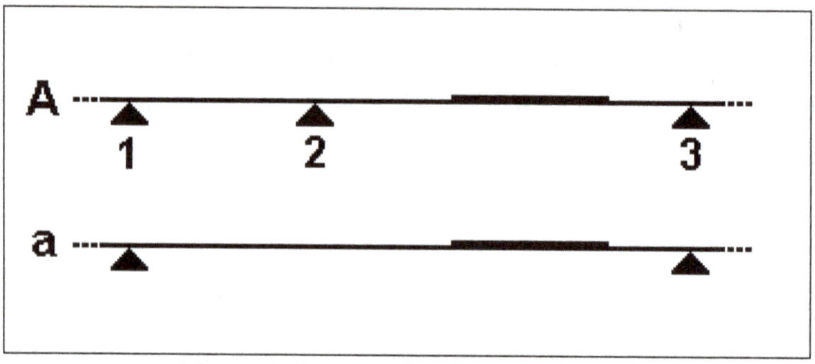

Schematic for RFLP by cleavage site loss.

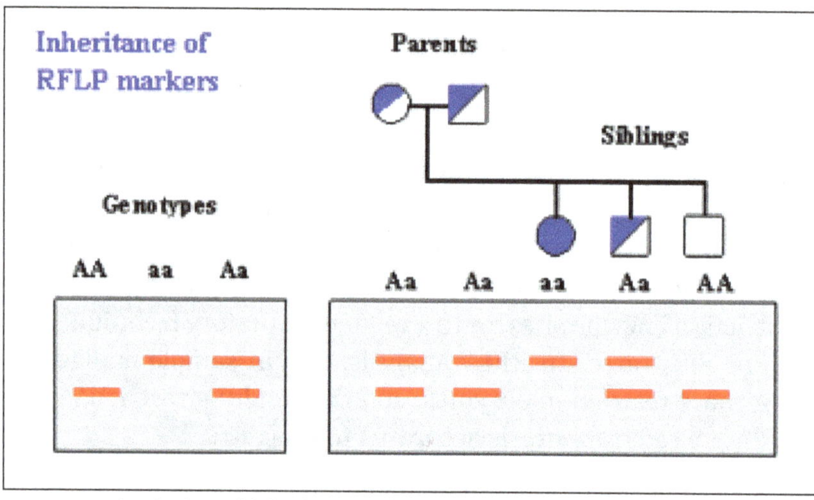

Analysis and inheritance of allelic RFLP fragments (NIH).

There are two common mechanisms by which the size of a particular restriction fragment can vary. In the first schematic, a small segment of the genome is being detected by a DNA probe (thicker line). In allele A, the genome is cleaved by a restriction enzyme at three nearby sites (triangles), but only the rightmost fragment will be detected by the probe. In allele a, restriction site 2 has been lost by a mutation, so the probe now detects the larger fused fragment running from sites 1 to 3. The second diagram shows how this fragment size variation would look on a Southern blot, and how each allele (two per individual) might be inherited in members of a family.

In the third schematic, the probe and restriction enzyme are chosen to detect a region of the genome that includes a variable number tandem repeat (VNTR) segment (boxes in schematic diagram). In allele c, there are five repeats in the VNTR, and the probe detects a longer fragment between the two restriction sites. In allele d, there are only two repeats in the VNTR, so the probe detects a shorter fragment between the same two restriction sites. Other genetic processes, such as insertions, deletions, translocations, and inversions, can also lead to polymorphisms. RFLP tests require much larger samples of DNA than do short tandem repeat (STR) tests.

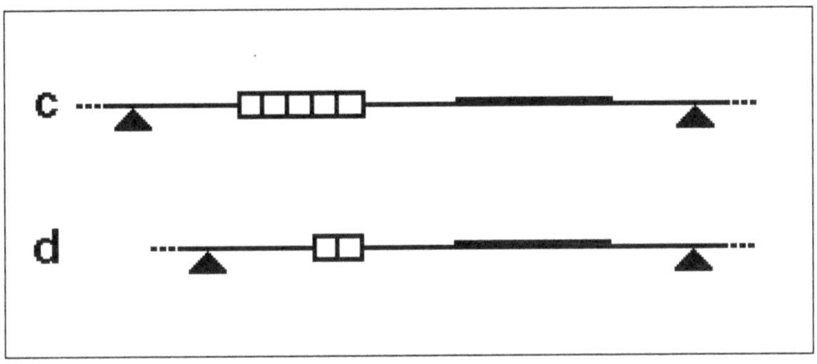

Schematic for RFLP by VNTR length variation.

Applications

Analysis of RFLP variation in genomes was formerly a vital tool in genome mapping and genetic disease analysis. If researchers were trying to initially determine the chromosomal location of a particular disease gene, they would analyze the DNA of members of a family afflicted by the disease, and look for RFLP alleles that show a similar pattern of inheritance as that of the disease. Once a disease gene was localized, RFLP analysis of other families could reveal who was at risk for the disease, or who was likely to be a carrier of the mutant genes. RFLP test is used in identification and differentiation of organisms by analyzing unique patterns in genome. It is also used in identification of recombination rate in the loci between restriction sites.

RFLP analysis was also the basis for early methods of genetic fingerprinting, useful in the identification of samples retrieved from crime scenes, in the determination of paternity, and in the characterization of genetic diversity or breeding patterns in animal populations.

Alternatives

The technique for RFLP analysis is, however, slow and cumbersome. It requires a large amount of sample DNA, and the combined process of probe labeling, DNA fragmentation, electrophoresis, blotting, hybridization, washing, and autoradiography can take up to a month to complete. A limited version of the RFLP method that used oligonucleotide probes was reported in 1985. The results of the Human Genome Project have largely replaced the need for RFLP mapping, and the identification of many single-nucleotide polymorphisms (SNPs) in that project (as well as the direct identification of many disease genes and mutations) has replaced the need for RFLP disease linkage analysis. The analysis of VNTR alleles continues, but is now usually performed by polymerase chain reaction (PCR) methods. For example, the standard protocols for DNA fingerprinting involve PCR analysis of panels of more than a dozen VNTRs.

RFLP is still used in marker-assisted selection. Terminal restriction fragment length polymorphism (TRFLP or sometimes T-RFLP) is a technique initially developed for characterizing bacterial communities in mixed-species samples. The technique has also been applied to other groups including soil fungi. TRFLP works by PCR amplification of DNA using primer pairs that have been labeled with fluorescent tags. The PCR products are then digested using RFLP enzymes and the resulting patterns visualized using a DNA sequencer. The results are analyzed either by simply counting and comparing bands or peaks in the TRFLP profile, or by matching bands from one or more TRFLP runs to a database of known species. The technique is similar in some aspects to temperature gradient or denaturing gradient gel electrophoresis (TGGE and DGGE).

The sequence changes directly involved with an RFLP can also be analyzed more quickly by PCR. Amplification can be directed across the altered restriction site, and the products digested with the restriction enzyme. This method has been called Cleaved Amplified Polymorphic Sequence (CAPS). Alternatively, the amplified segment can be analyzed by allele-specific oligonucleotide (ASO) probes, a process that can often be done by a simple dot blot.

Applications of RFLP Marker in Plants

Linkage Construction

Restriction fragment length polymorphism facilitated in the construction of linkage maps of any genotypes and genes linked to certain specific agronom characters have been successfully mapped by RFLP markers. In conventional method, genetic mapping is generally accomplished during crossing over recombination process by F_1 plant.

The action of genes after looking at the flower colour, plant height etc. and those methods are considered to be time consuming. Besides, the inheritance pattern of RFLP marker, segregate exactly and follow mendelian rule. Construction of RFLP genetic map involves series of steps. After the selection of a parent plant by wide range of survey DNA is isolated and digested with restriction enzymes and screened for polymorphism.

In the next step, production of mapping population is accomplished by crossing selected parent plant to produce F_1 plants. Consequently F_1 plants are selfed to produce F_2, which can be screened for segregation of RFLP or back crossed to one of the parent.

This would pave the speed up process without being crossed in the final step. Once population mapping is procured, DNA is extracted from each individual plant in the population. It can be presumed that the chromosome of each plant in the population contains wide array of parental chromosome segments.

Mapping population is followed by scoring RFLP in the individual plants of mapping population. The data obtained from the scoring of mapping population and information from the library is used to construct linkage map.

Disease Management

RFLP technique has been implicated in disease management by detection of plant-pathogen. RFLP analysis has been used to distinguish Sclerotinia species. Probes are obtained from nuclear and mitochondrial genes including ribosomal RNA genes. Probe from Neurospora crassa which helps in distinguishing of two or seven species of Sclerotinia.

RFLP based on VNTR from genomic library of Pithium has been used to identify plant pathogen, Pithium. RFLP marker helps in the transfer of certain fruit quality trait from wild into cultivated crops.

Tightly linked RFLP marker for sucrose accumulator gene has been transferred into tomato (Lycopersicum esculentum) from wild tomato (Lycopersicum chmielewski). RFLP marker has been identified for genes conferring resistance to the root cyst nematode in potato.

Bentolina identified tight linkage of a RFLP marker to the Ht1 gene of maize that confer resistance to the fungal pathogen Helminthosporium turcicum by four pairs of near isogenic lines. Genetic dwarfism caused by gene has been mapped with the help of RFLP markers on the long arm of chromosome in case of barley. RFLP is used to investigate variation in mitochondrial genes.

Analysis of Somaclonal Variation

Genetic variation can be induced by invitro culture which is otherwise known as somaclonal variation. Tissue culture derived plants are prone to genetic alteration or deliberatively by induced in vitro. RFLP molecular markers are potential tool in the detection of genetic variation.

Extent of genetic variation is required to know before the introduction of single gene into the plant. If a degree of genetic variation is high then it is useful tool for plant breeders rather than for commercial release.

Prediction of Genetic Distance

Increased performance of hybrid crops heterosis is due to the genetic distance between the parent plants used in cross. Molecular markers like RFLP can be efficiently used to estimate genetic distance between possible parents.

Assessment of genetic diversity in plant genetic resource collections provides better management when space and resources are constraints. RFLP molecular method can be employed to distinguish short day onions from long day (which is otherwise difficult to identify). Identification and preservation of short day onions is indispensable because of its high genetic diversity in nature.

AFLP

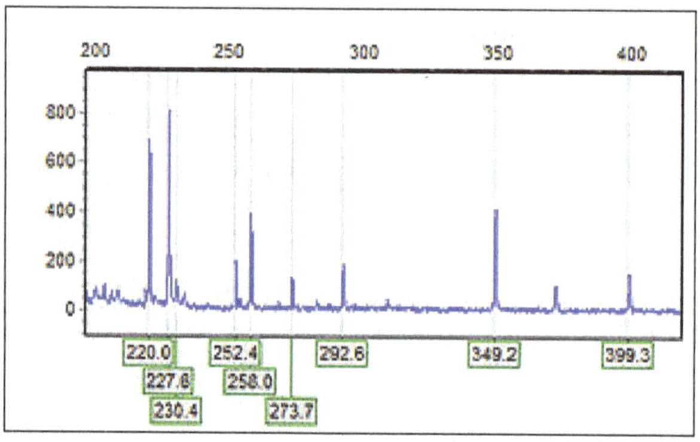

Example of AFLP data from a capillary electrophoresis instrument.

AFLP-PCR or just AFLP is a PCR-based tool used in genetics research, DNA fingerprinting, and in the practice of genetic engineering. Developed in the early 1990s by Keygene, AFLP uses restriction enzymes to digest genomic DNA, followed by ligation of adaptors to the sticky ends of the restriction fragments. A subset of the restriction fragments is then selected to be amplified. This selection is achieved by using primers complementary to the adaptor sequence, the restriction site sequence and a few nucleotides inside the restriction site fragments. The amplified fragments are separated and visualized on denaturing on agarose gel electrophoresis, either through autoradiography or fluorescence methodologies, or via automated capillary sequencing instruments.

Although AFLP should not be used as an acronym it is commonly referred to as "Amplified fragment length polymorphism", the resulting data are not scored as length polymorphisms, but instead as presence-absence polymorphisms.

AFLP-PCR is a highly sensitive method for detecting polymorphisms in DNA. The technique was originally described by Vos and Zabeau in 1993. In detail, the procedure of this technique is divided into three steps:

- Digestion of total cellular DNA with one or more restriction enzymes and ligation of restriction half-site specific adaptors to all restriction fragments.
- Selective amplification of some of these fragments with two PCR primers that have corresponding adaptor and restriction site specific sequences.

- Electrophoretic separation of amplicons on a gel matrix, followed by visualisation of the band pattern.

Applications

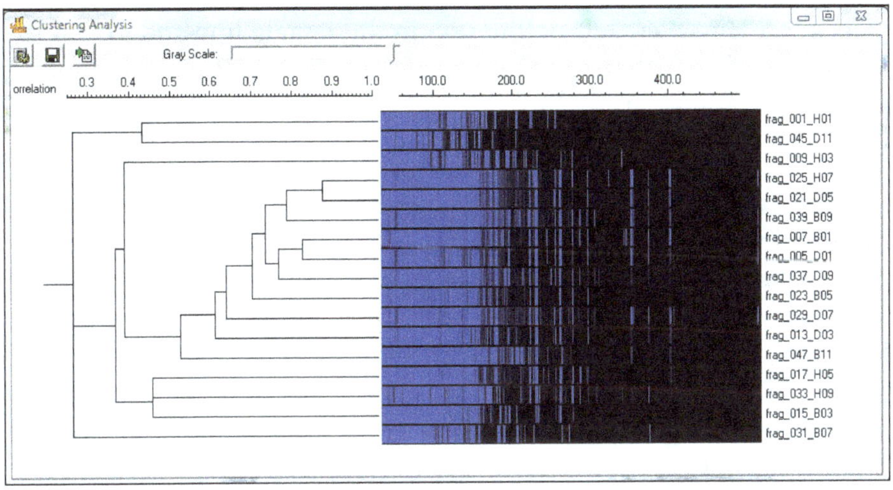

AFLP phylogeny analysis using a dendrogram.

The AFLP technology has the capability to detect various polymorphisms in different genomic regions simultaneously. It is also highly sensitive and reproducible. As a result, AFLP has become widely used for the identification of genetic variation in strains or closely related species of plants, fungi, animals, and bacteria. The AFLP technology has been used in criminal and paternity tests, also to determine slight differences within populations, and in linkage studies to generate maps for quantitative trait locus (QTL) analysis.

There are many advantages to AFLP when compared to other marker technologies including randomly amplified polymorphic DNA (RAPD), restriction fragment length polymorphism (RFLP), and microsatellites. AFLP not only has higher reproducibility, resolution, and sensitivity at the whole genome level compared to other techniques, but it also has the capability to amplify between 50 and 100 fragments at one time. In addition, no prior sequence information is needed for amplification. As a result, AFLP has become extremely beneficial in the study of taxa including bacteria, fungi, and plants, where much is still unknown about the genomic makeup of various organisms.

The AFLP technology is covered by patents and patent applications of Keygene N.V. AFLP is a registered trademark of Keygene N.V.

Applications of AFLP Marker in Plant Breeding

Variety identification, phylogeny and diversity in plants AFLP analysis is a multilocus PCR technology that generates a considerably amount of locus-specific bands, a high percentage of which can be polymorphic. Estimates of genetic distances based on differences in AFLP patterns are informative about genetic diversity, phylogeny and the geographic origins of genotypes and gene pools of plants.

F_1 hybrids are the result of a cross between a female and male parental homozygous breeding line and often have agronomic performance superior to those of the parental homozygous lines. However, self-pollination in the female line and pollen from other male lines may interfere with the production of F^1 seeds. AFLP analysis allows the identification of the contaminating variety.

In a study on rice, Mackill 24 compared the levels of polymorphism for AFLP, RAPD and microsatellite markers on 12 japonica and two indica rice cultivars. Using 18 AFLP primer pairs, 143 of a total of 539 generated bands were polymorphic. Twenty-one RAPD primers produced 103 bands, 43 of which were polymorphic. The number of alleles detected using 14 microsatellite primers ranged from one to six. Although the rice accessions were classified into the same subspecies using all marker types, the level of polymorphism varied considerably. Within japonica cultivars, the average percent polymorphism between any two accessions was 22% for AFLP, 24% for RAPD and 36% for microsatellite markers (monomorphic bands excluded). The average percent polymorphism between indica and japonica accessions was highest for microsatellite (76%) and AFLP (65%) markers as compared with RAPD (35%) markers.

Although microsatellite markers produced the highest level of polymorphism, the total number of polymorphic bands was much higher for AFLPs with an average of eight per gel. Only one locus per gel can be scored for microsatellite analysis unless multiplexing is employed. To evaluate the genetic mapping of AFLP markers, 80 F^2 plants of an indica japonica cross were analyzed using seven primer combinations. An existing RFLP-RAPD map containing 115 RFLP, 10 RAPD markers and one morphological trait were used to map 50 of 54 scored AFLP bands to specific chromosomes. The markers were distributed throughout the rice genome revealing the potential of AFLP markers for mapping important genes in rice.

Plant Breeding

Despite some apparent clustering AFLP markers (i.e., discriminatory fragments by AFLP analysis) cover all chromosomes and are inherited in a Mendelian way, which is a prerequisite for molecular marker applications in breeding analysis. The AFLP technology has been successfully used in several applications of marker-assisted breeding. One of them is the selection of the offspring of a cross in order to have, for example, the lowest percentage of donor-specific alleles. In some breeding programs interesting characters are introduced into an elite (or commercial) line by crossing it with the donor of such character. After several backcrosses of the offspring with the elite material, most of the commercial genome with interesting traits, is recovered. Vantoai observed that using AFLPs in soybean to select the offspring with the lowest percentage of donor genome, only six generations were needed to reduce the contribution of the donor parent to an average of 1%. In addition, AFLP has practical applications for plant DNA genetics. It has been successfully used for marker-assisted plant breeding, for construction of high-density molecular linkage maps, and for positional cloning of genes of interest.

Indirect Selection of Agronomically Important Properties

For monogenic traits, AFLP analysis needs a large collection of primer pairs to yield markers closely linked to these traits for diagnostic use. This technique allowed the localization of genes in plants that confer resistance to viruses, nematodes, fungi or bacteria and positional cloning of the relevant genes.

For the localization of polygenic quantitative traits AFLP analysis allows the fast and efficient construction of dense genetic maps. Genotyping of large segregating populations can lead to the localization of one or more quantitative trait loci.

The DNA markers generated in the study by Meksem will aid in the further characterization of the R1 gene and may help to elucidate the mechanism of race-specific resistance to P. infestans. A high-resolution map of the Gro1 locus on potato Chromosome VII was constructed by Ballvora using RFLP, RAPD and AFLP markers. The dominant allele, Gro1, confers resistance to a root cyst nematode, Globodera rostochiensis. AFLP patterns were generated using 24 primers for a total of 144 primer combinations tested. Bulked segregant analysis was used to select RAPD and AFLP markers closely linked to Gro1 in a segregating population of 1105 plants. Approximately 3000–3300 informative loci were analyzed. Three RFLP, one RAPD and two AFLP markers were linked to the Gro1 locus. The RFLP and RAPD markers were inseparable from the Gro1 locus, while the two AFLP markers flanked Gro1 at genetic distances of 0.6 cM and 0.8 cM (where 1 cM = 100 kilo base pairs (kbp).

The closely linked AFLP markers identified will facilitate molecular cloning of the Gro1 gene based on map position. Cloning the Gro1 gene may contribute to the elucidation of the mechanism of resistance, and the Gro1 gene can be directly transformed into susceptible potato cultivars that are agronomically sound. Using bulked segregant pools from F_2 progeny, of the cross Lycopersicon esculentum (Cf-9) x L. pennellii and AFLP, Thomas 36 screened approximately 42,000 AFLP loci for tight linkage to the tomato Cf-9 gene. The Cf-9 gene, on Chromosome 1 of tomato, codes for resistance to the leaf mould pathogen, Cladosporium fulvum. A total of 728 primer combinations were tested using 13 PstI and 56 MseI primers. Of the 42,000 AFLP loci inspected for polymorphism, approximately 14,000 were potentially informative. This equates to one AFLP marker every 70 kbp if the markers are distributed at random over the tomato genome estimated at 109 bp. Analysis of F_2 recombinants identified three markers co-segregating with Cf-9. Further analysis of plasmid clones containing the gene allowed the position of the markers to be determined.

Two of the markers were found to be located on opposite sides of the Cf-9 gene, separated by 15.5 kbp. The AFLP technique was used by Cnops in a chromosome landing strategy, for the isolation of the TRN1 gene. The tornado1 (trn1) mutation was identified in a T-DNA transgenic line of Arabidopsis (ecotype C24) and is located 5 cM from a T-DNA insertion. Visible markers and RFLP markers were used to map the TRN1 locus to the bottom half of Chromosome 5.

Using Populus spp. as a model tree, Cervera showed and discussed three possible applications of AFLPTM in forest tree breeding. The first application, which is described in more detail by Cervera involves the use of AFLP to identify molecular markers tightly linked to the Mer locus conferring resistance against three races (E1, E2 and E3) of the leaf rust fungus, Melampsora larici-populina. This disease causes premature defoliation, leaving the tree susceptible to secondary pathogen infection and environmental stress, and can reduce growth by more than 20%. The study was carried out using a hybrid progeny from a cross between a resistant P. deltoides female and a susceptible P. nigra male. Data suggested that resistance is determined by a single dominant marker. AFLP analysis using 144 primer combinations and bulked segregant analysis identified three AFLP markers tightly linked to the Mer locus.

These markers can be used in current breeding programs and for future cloning of the resistance gene. The second application involves the generation of genomic fingerprints for each species of Populus, using AFLP, for identification and taxonomic analysis. The results for the study on the diversity among Populus spp. have not yet been reported. The final application involves the generation of high-density linkage maps of P. deltoides, P. nigra and P. trichocarpa using the 'two-way pseudo-testcross' strategy in combination with AFLP analysis. The 'two-way pseudotestcross' mapping strategy is based on linkage analysis of those markers that are heterozygous in one parent and null in the other parent. These results have also not yet been published. The multiple studies on Populus spp. using AFLP reveals the versatility of the technique for a variety of applications.

Brigneti combined AFLP and bulked segregant analysis of an F_1 tetraploid potato population to identify AFLP markers tightly linked to the Rysto gene conferring resistance to all strains of the potato virus Y. Rysto is a dominant gene and was mapped to Chromosome XI of the potato genome. Preliminary AFLP analysis using 216 primer combinations identified three markers linked to and on either side of Rysto. Four additional AFLP markers were identified in a population of 360 segregating progeny of a potato cross between a resistant (Rysto) and a susceptible parent. Two markers co-segregated with Rysto and the other two were located on either side of Rysto, separated by one recombination event. AFLP analysis was successfully used to map the Rysto gene, however, the closest flanking markers were estimated to be 0.6 cM apart (where 1 cM =1 Mb). Additional markers around the resistance gene are necessary for positional cloning.

High-density Molecular Linkage Maps

Generation of highresolution genetic maps in plants in a relatively well-characterized mapping population of potato, Van Eck studied the localization of AFLP markers relative to a mapping population of 197 RFLP, nine isoenzyme, and 11 morphological markers. Six primer combinations generated 264 segregating AFLP amplification products in a diploid backcross population from non-inbred potato parents. The segregating patterns of the AFLP amplification products observed in the offspring were used to evaluate the inheritance of putative AFLP loci. AFLP mapping generated more than double the number of genetic markers when compared to RFLP.

Although the AFLP markers were randomly distributed, they targeted regions already mapped by RFLP markers such that the total map length increased only 5% from 1120 to 1170 centimorgans (cM). Van Eck concluded that AFLP might substitute for other marker systems. The AFLP amplification products mapped by Van Eck will enable the chromosomal identity of linkage groups to be established in future mapping studies because of the locus specificity of AFLP markers. In addition, the identification and mapping of comigrating AFLP fragments from other genomes will be facilitated. In a similar study on barley, Becker generated 118 AFLP markers using 16 primer combinations. Two of the linked AFLP markers could not be assigned to one of seven linkage groups. The remaining 116 markers were mapped to existing barley RFLP map including five microsatellite and four protein marker loci.

AFLP and RFLP markers were used by Meksem to construct a high-resolution genetic map of the R1 locus on Chromosome V of potato. The R1 allele confers, on potato, a racespecific resistance to Phytophthora infestans, a fungal pathogen. Bulked segregant analysis and AFLP markers were used to select markers mapping less than 1 cM from the R1 gene. Of 3200 informative AFLP loci, 29 displayed linkage to the R1 locus. These 29 fragments were amplified by 21 of a total of 108 primer combinations tested. Genotypic analysis of 461 F1 plants revealed that eight loci mapped within a genetic distance of 6 cM as defined by two marker loci, GP21 and GP179, flanking the R1 locus. Only six were linked in coupling with the resistance allele R1.

On the other hand, the AFLP technique was applied to recombinant classes, within a 3-cM region around the TRN1 locus, to build a high-resolution map in this region. Two different restriction enzymes, SacI and EcoRI, were used in combination with MseI, to generate template DNA for amplification. Approximately 26,000 AFLP fragments were generated using 192 SacI-MseI primer combinations and 112 EcoRI-MseI primer combinations. Seventeen AFLP markers were identified, four of which were so closely linked to the TRN1 locus that no recombination event was detected. Three AFLP fragments were purified from the dried gel and cloned: two trn1 co-segregating markers and one marker located immediately below trn1.

To facilitate the recovery of the fragments, a simplified banding pattern (with 16-fold fewer bands) was generated using SacI and MseI primers with an additional selective nucleotide on the 3′-ends.

The linkage map of sugar beet (Beta vulgaris L.) was extended by Schondelmaier using the AFLP technique on an F_2 population consisting of 94 individuals. A total of 120 AFLP markers generated using four primer combinations were integrated into an existing linkage map based on 207 RFLP loci. The AFLP markers were evenly distributed over all nine linkage groups with the exception of Linkage Group V. The low number of AFLP loci on Linkage Group V was thought to be due to the low number of primer combinations tested. Integration of AFLP loci served to extend the sugar beet map from 544 cM to 557 cM. The increase in genetic length was brought about through the addition of AFLP markers at the ends of Chromosomes II, IV and VII, as well as AFLP loci filling some gaps in the RFLP map. AFLP technology proved to be a promising tool for estimating the genetic diversity of sugar beet breeding lines because of the even distribution of AFLP markers throughout the genome.

SNP Markers and their Impact on Plant Breeding

Allelic variations within a genome of the same species can be classified into three major groups that include differences in the number of tandem repeats at a particular locus microsatellites, or simple sequence repeats (SSRs), segmental insertions/deletions (InDels), and single nucleotide polymorphisms (SNPs). In order to detect and track these variations in the individuals of a progeny at DNA level, researchers have been developing and using genetic tools called molecular markers. Although SSRs, InDels, and SNPs are the three major allelic variations discovered so far, a plethora of molecular markers were developed to detect the polymorphisms that resulted from these three types of variation. Evolution of molecular markers has been primarily driven by the throughput and cost of detection method and the level of reproducibility. Depending on detection method and throughput, all molecular markers can be divided into three major groups: (1) low-throughput, hybridization-based markers such as restriction fragment length polymorphisms (RFLPs); (2) medium-throughput, PCR-based markers that include random amplification of polymorphic DNA (RAPD), amplified fragment length polymorphism (AFLP), SSRs; (3) high-throughput (HTP) sequence-based markers: SNPs. In late eighties, RFLPs were the most popular molecular markers that were widely used in plant molecular genetics because they were reproducible and codominant. However, the detection of RFLPs was an expensive, labor- and time-consuming process, which made these markers eventually obsolete. Moreover, RFLP markers were not amenable to automation. Invention of PCR technology and the application of this method for the rapid detection of polymorphisms overthrew low-throughput RFLP markers, and new generation of PCR-based markers emerged in the beginning of nineties. RAPD, AFLP, and SSR markers are the major PCR-based markers that research community has been using in various plant systems. RAPDs are able to simultaneously detect polymorphic loci in various regions of a genome. However, they are anonymous and the level of their reproducibility is very low due to the non-specific binding of short, random primers.

Although AFLPs are anonymous too, the level of their reproducibility and sensitivity is very high owing to the longer +1 and +3 selective primers and the presence of discriminatory nucleotides at 3 end of each primer. That is why AFLP markers are still popular in molecular genetics research in crops with little to zero reference genome sequence available. However, AFLP markers did not find widespread application in molecular breeding owing to the lengthy and laborious detection method, which was not amenable to automation either. Therefore, it was not surprising that soon after the discovery of SSR markers in the genome of a plant, they were declared as "markers of choice", because SSRs were able to eliminate all drawbacks of the above-mentioned DNA marker technologies. SSRs were no longer anonymous; they were highly reproducible, highly polymorphic, and amenable to automation. Despite the cost of detection remaining high, SSR markers had pervaded all areas of plant molecular genetics and breeding in late 90s and the beginning of 21st century. However, during the last five years, the hegemony of medium-throughput SSRs was eventually broken by SNP markers. First discovered in human genome, SNPs proved to be universal as well as the most abundant forms of genetic variation among individuals of the same species. Although SNPs are less polymorphic than SSR markers because of their biallelic nature, they easily compensate this drawback by being abundant, ubiquitous, and amenable to high- and ultra-high-throughput automation. However, despite these obvious advantages, there were only a limited number of examples of application of SNP markers in plant breeding by 2009.

SNP Discovery in Complex Plant Genomes

While SNP discovery in crops with simple genomes is a relatively straightforward process, complex genomes pose serious obstacles for the researchers interested in developing SNPs. One of the major problems is the highly repetitive nature of the plant genomes. Prior to the emergence of next-generation sequencing (NGS) technologies, researchers used to rely on different experimental strategies to avoid repetitive portions of the genome. These include discovery of SNPs experimentally by resequencing of unigene-derived amplicons using Sanger's method and in silico SNP discovery through the mining of SNPs within EST databases followed by PCR-based validation. Although these approaches allowed the detection of gene-based SNPs, their frequency is generally low in conserved genic regions, and they were unable to discover SNPs located in low-copy noncoding regions and intergenic spaces. Additionally, amplicon resequencing was an expensive and labor-intensive procedure. As many crops are ancient tetraploids with mosaics of scattered duplicated regions, in silico and experimental mining of EST databases resulted in the discovery of a large number of nonallelic SNPs that represented paralogous sequences and were suboptimal for application in molecular breeding. Recent emergence of NGS technologies such as 454 Life Sciences, HiSeq, SOLiD and Ion Torrent has eliminated the problems associated with low throughput and high cost of SNP discovery. Transcriptome resequencing using NGS technologies allows rapid and inexpensive SNP discovery within genes and avoids highly repetitive regions of a genome. This methodology was successfully applied in several plant genomes, including maize, canola, eucalyptus, sugarcane, tree species, wheat, avocado, and black currant. Originally developed for human disease diagnostic research, the NimbleGen sequence capture technology (Roche Applied Science, IN) brought the detection of gene-based SNPs in plants into higher throughput and coverage level. This technology consists of exon sequence capture and enrichment by microarray followed by NGS for targeted resequencing. Similar in-solution target capture technologies, such as Agilent SureSelect, are also commercially available for genome/exome mining studies. However, this technology would be efficient only for crops with available reference genome sequence or large transcriptome (EST) datasets, since the design of capture probes requires these reference resources.

Despite the attractiveness of SNP discovery via transcriptome or exome resequencing, this process is targeted, focusing solely on coding regions. It is obvious that the availability of SNPs within coding sequences is a very powerful tool for molecular geneticists to detect a causative mutation. However, often QTL are located in noncoding regulatory sequences such as enhancers or locus control regions, which could be located several megabases away from genes within intergenic spaces. Discovery of SNPs located within those regulatory elements via transcriptome or exon sequencing is limited. In order to discover SNPs in a genome-wide fashion and avoid repetitive and duplicated DNA, it is very important to employ genome complexity reduction techniques coupled with NGS technologies. Several genome complexity reduction techniques have been developed over the years, including High Cot selection, methylation filtering, and microarray-based genomic selection. These techniques mainly reduce the number of repetitive sequences but lack the power to recognize and eliminate duplicated sequences, which cause the detection of false-positive SNPs. Unlike the above-mentioned techniques, recently developed genome complexity reduction technologies such as Complexity Reduction of Polymorphic Sequences (CRoPS) and Restriction Site Associated DNA (RAD)are computationally well equipped and capable of filtering out duplicated SNPs. These systems were successfully applied to discover SNPs in crops with and without reference genome sequences.

Although several complexity reduction approaches are being developed to generate data from NGS platforms, it is often challenging to identify candidate SNPs in polyploid crops species such as potato, tobacco, cotton, canola, and wheat. In general, minor allele frequency could be used as a measure to identify candidate SNPs in diploid species. However, in polyploid crops, you often find loci that are polymorphic within a single genotype due to the presence of either homoeologous loci from the individual subgenomes (homoeologous SNPs) or paralogous loci from duplicated regions of the genome. Such false positive SNPs are not useful for genetic mapping purposes and often lead to a lower validation rate during assays. Successful SNP validation in allopolyploids depends upon differentiation of the sequence variation classes. Use of haplotype information beside the allelic frequency would help to identify homologous SNPs (true

SNPs) from those of homoeologous loci (false positives). Bioinformatic programs such as HaploSNPer would facilitate identification of candidate loci for assay design purposes in polyploid crops. Elimination of homoeologous loci for the assay design process would improve the validation rate. Such approaches could also be extended to other complex and highly repetitive diploid genomes such as barley. Complexity reduction approaches, combined with sophisticated computational tools, would expedite SNP discovery and validation efforts in polyploids.

Although CRoPS and RAD technologies are powerful tools to detect SNPs in genome-wide fashion, they can hardly be called HTP, because on an average only ~1,000 SNPs pass stringent quality control. While these numbers are enough to generate genetic linkage maps of reasonable saturation and carry out preliminary QTL mapping, they are not adequate to implement genome-wide association studies (GWAS). Depending on the rate of linkage disequilibrium decay, GWAS might require several million genetic landmarks. From this point of view, genotyping by-sequencing (GBS) technique offers many more opportunities. Discovery of a large number of SNPs using GBS was demonstrated in maize and sorghum. GBS not only increases the sequencing throughput by several orders of magnitude but also has multiplexing capabilities. To eliminate a large portion of repetitive sequences, a type II restriction endonuclease, ApeKI, is applied to digest DNA prior to sequencing to generate reduced representation libraries (genome complexity reduction component), which are further subject to sequencing. In polyploid crops, GBS might be challenging, but the associated complexity reduction methods could be used for SNP discovery. For discovery purposes, the availability of a reference genome is not an absolute requirement to implement GBS approach. However, in organisms that do not have a reference genome, GBS-derived SNPs must be validated using one of the techniques which might dramatically increase per marker price. Validation needs to be done primarily to discard paralogous SNPs. For organisms with a reference genome sequence, the validation step is replaced by in silico mapping of the sequenced fragments to the genome. Although GBS has the potential to discover several million SNPs, one of the major drawbacks of this technique is large numbers of missing data. To solve this problem, computational biologists developed data imputation models such as BEAGLE v3.0.2 and IMPUTE v2, to bring imputed data as close as possible to the real data.

SNP Validation and Modern Genotyping Platforms and Chemistries

The availability of reference sequence and sophisticated software does not always guarantee that the discovered SNP can be converted into a valid marker. In order to insure that the discovered SNP is a Mendelian locus, it has to be validated. The validation of a marker is the process of designing an assay based on the discovered polymorphism and then genotyping a panel of diverse germplasm and segregating population. Compared to the collection of unrelated lines, a segregating population is more informative as a validation panel because it allows the inspection of the discriminatory ability and segregation patterns of a marker which helps the researcher to understand whether it is a Mendelian locus or a duplicated/repetitive sequence that escaped the software filter.

The most popular HTP assays/chemistries and genotyping platforms that are currently being used for SNP validation are Illumina's BeadArray technology-based Golden Gate (GG) and Infinium assays, Life Technologies' TaqMan assay coupled with OpenArray platform (TaqMan OpenArray Genotyping system, Product bulletin), and KBiosciences' Competitive Allele Specific PCR (KASPar) combined with the SNP Line platform. These modern genotyping assays and platforms differ from each other in their chemistry, cost, and throughput of samples to genotype and number of SNPs to validate. The choice of chemistry and genotyping platform depends on many factors that include the length of SNP context sequence, overall number of SNPs to genotype, and finally the funds available to the researcher, because most of these chemistries still remain cost intensive. Comparative analyses of these four genotyping assays and platforms were described in Kumpatla

Though all genotyping chemistries and platforms are applicable to generate genotypic data in polyploid crops, analysis of SNP calls is somewhat challenging in polyploids due to multiallele combinations in the genotypes. SNPs in polyploid species can be broadly classified as simple SNPs, hemi-SNPs, and homoeo-SNPs. Here, we describe simple, hemi-, and homoeo-SNPs using an example of allele calls in tetraploid and diploid cotton species. Genomes of tetraploid cotton species, Gossypium hirsutum (AD_1) and G. barbadense (AD_2), consist of two subgenomes A and D, where A genome was derived from diploid progenitors, such as G. herbaceum (A_1) and G. arboreum (A_2), and D genome resulted from another diploid progenitor G. raimondii (D_5). Simple, or true SNPs are markers that detect allelic variation between homologous loci of the same subgenome of two tetraploid samples. For example, in Figure, a SNP marker clearly detects polymorphism within A subgenomes of G. hirsutum (AD_1) and G. barbadense (AD_2) and separates samples into homozygous A (blue) and B (red) clusters. This marker does not discriminate polymorphism in D subgenome, because the D genome allele is absent there (pink dot in G. raimondii). In contrast to simple SNPs, hemiSNPs detect allelic variation in the homozygous state in one sample and the heterozygous state in the other sample.

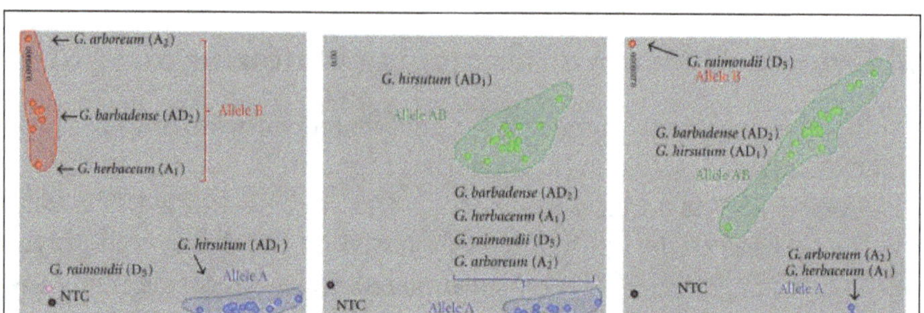

Segregation patterns of simple, hemi-, and homoeo-SNPs assayed using KASPar

chemistry across tetraploid cotton species G. hirsutum (AD_1) and G. barbadense (AD_2) and their diploid progenitors [G. arboreum, G. herbaceum (A subgenome), and G. raimondii (D-subgenome)]. (a) Simple, or true SNP detects allelic variation between homologous loci of A subgenome of G. hirsutum (AD_1) and G. barbadense (AD_2). (b) Hemi-SNPs detect allelic variation in homozygous state in G. barbadense (AD_2) and heterozygous state in G. hirsutum (AD_1). (c) Homoeo-SNP detects homoelogous and paralogous loci both in A and D subgenomes, which are monomorphic between G. barbadense (AD_2) and G. hirsutum (AD_1). Allele calls depicted in blue, red, green, pink, and black represent alleles A, B, and AB, no amplification or missing locus, and no template control (NTC), respectively.

In figure, SNP marker detects both alleles (A and B) in G. hirsutum (heterozygous green cluster) and one allele A in G. barbadense (a homozygous blue cluster) and could be vice versa. Homoeo-SNPs detect homoeologous and possibly paralogous loci both in A and D subgenomes and result in monomorphic loci in tetraploid species. In figure A genome progenitors (G. herbaceum and G. arboreum) had allele A (blue) and D genome progenitor (G. raimondii) had allele B (red), but both tetraploid species (G. hirsutum and G. barbadense) were grouped into heterozygous AB (green) cluster. As homoeo-SNPs can detect paralogous loci, the diploid progenitors both have different alleles.

Simple SNPs as well as hemi-SNPs are useful markers for genetic mapping and diversity screening studies. Simple SNPs segregate like the markers in diploids in most of the mapping populations and would account for approximately 10–30% of total polymorphic SNPs in various polyploid crop species. Hemi-SNPs form a major category (30–60%) of polymorphic SNPs in a polyploid crop species and could be used for genetic mapping purposes in F2, RIL, and DH populations. Homoeo-SNPs are of lesser value for mapping purposes as most of the genotypes result in heterologous loci due to polymorphism between the homoeologous genomes or duplicated loci within each of the polyploid genotypes.

Application of SNP Markers in Gene/QTL Discovery

Biparental Approach

Genetic mapping studies involve genetic linkage analysis, which is based on the concept of genetic recombination during meiosis. This encompasses developing genetic linkage maps following genotyping of individuals in segregating populations with DNA markers covering the genome of that organism. Since their discovery in the 1980s, DNA-based markers have been widely used in developing saturated genetic linkage maps as well as for the mapping and discovery of genes/QTL. With the large-scale availability of the sequence information and development of HTP technologies for SNP genotyping, SNP markers have been increasingly used for QTL mapping studies. This is primarily, because SNPs are highly abundant in the genomes and, therefore, they can provide the highest map resolution compared to other marker systems. A review of the selected examples of QTL and gene discovery using SNP markers is presented.

Examples in Rice

A recent study on QTL analysis in rice for yield and three-yield-component traits, number of tillers per plant, number of grains per panicle, and grain weight compared a SNP-based map to that of a previous RFLP/SSR-based QTL map generated using the same mapping population. Using the ultra-high-density SNP map, the authors showed that this map had more power and resolution relative to the RFLP/SSR map. This was clearly evident by the analysis of the two main QTL for grain weight, kgw3a (GS3) and kgw5 (GW5/qSW5). Using the SNP bin map, GW5/qSW5 QTL for grain width was accurately narrowed down to a 123 kb region as compared to the 12.4 Mb region based on the RFLP/SSR genetic map. Likewise, GS3 QTL for grain length was mapped to a 197 kb interval in comparison to 6 Mb region with the RFLP/SSR genetic map. Beside the power and the resolution, maps based on high-density SNP markers are also highly suitable for fine mapping and cloning of QTL and at times SNPs on these maps are also functionally associated with the natural variation in the trait. In another QTL mapping project, SNP and InDel markers were used to fine map qSH1 gene, a major QTL of seed shattering trait in rice. The QTL were initially detected using RFLP and RAPD markers on F_2 plants. Using large BC_4F_2 and BC_3F_2 populations in fine mapping approach with SNP and InDel markers. They further showed that this SNP in the 5 regulatory region of the qSH1 gene caused loss of seed shattering. Fine mapping approach was also taken to positionally clone the rice bacterial blight resistance gene xa5, by isolating the recombination breakpoints to a pair of SNPs followed by sequencing of the corresponding 5 kb region. Several studies have shown that the SNPs and InDels are highly abundant and present throughout the genome in various species including plants. SNP genotyping is a valuable tool for gene mapping, map-based cloning, and marker assisted selection (MAS) in crops. A study was conducted to assess the feasibility of SNPs and InDels as DNA markers in genetic analysis and marker-assisted breeding in rice by analyzing these sequence polymorphisms in the genomic region containing Piz and Piz-t rice blast resistance genes and developing PCR-based SNP markers. SNPs were abundant in the Piz and Piz-t (averaging one SNP every 248 bp), while InDels were much lower. This dense distribution of SNPs helped in developing SNP markers in the vicinity of these genes. Advancements in rice genomics have led to mapping and cloning of several genes and QTL controlling agronomically important traits, enabled routine use of SNP markers for MAS, gene pyramiding, and marker-assisted breeding (MAB).

Examples in Maize

SNP markers have facilitated the dissection of complex traits such as flowering time in maize. Using a set of 5000 RILs, which represent the nested association mapping (NAM) population and genotyping with 1,200 SNP markers, the genetic architecture of flowering time is controlled by small additive QTL rather than a single large-effect

QTL. The same NAM population was used for mapping resistance to northern leaf blight disease. Twenty-nine QTL were discovered and candidate genes were identified with genomewide NAM approach using 1.6 million SNPs. Proprietary SNP markers developed by companies are being predominantly used in their private breeding programs. A study from Pioneer Hi-Bred International Inc. reported identifying a high-oil QTL (qHO6) affecting maize seed oil and oleic acid contents. This QTL encodes an acyl-CoA:diacylglycerol acyltransferase (DGAT1-2), which catalyzes the final step of oil synthesis.

Examples in Wheat

Recent advances in wheat genomics have led to the implementation of high-density SNP genotyping in wheat. Gene-based SNP markers were developed for Lr34/Yr18/Pm38 locus that confers resistance to leaf rust, stripe rust, and powdery mildew diseases. These markers serve as efficient tools for MAS and MAB of disease resistant wheat lines. Another economically important wheat disease, Fusarium head blight (FHB), has been extensively studied. Several QTL controlling FHB resistance have been identified, with the most important being Fhb1. Recently, SNP markers were mapped between the known flanking markers for Fhb1. These new markers would be useful for MAS and fine mapping towards cloning the Fhb1 gene. MAS in wheat has been extensively applied for simple traits that are difficult to score.

Examples in Soybean

In order to improve the effectiveness of MAS and clone soybean aphid resistance gene, Rag1, fine mapping was done to accurately position the gene, which was previously mapped to a 12 cM interval. Another aphid resistance gene, Rag2, originally mapped to a 10 cM interval, was fine mapped to a 54 kb interval using SNP markers that were developed by resequencing of target intervals and sequence-tagged sites. In another study that used a similar approach, SNP markers tightly linked to a QTL conferring resistance to southern root-knot nematode by developing these SNP markers from the bacterial artificial chromosome (BAC) ends and SSR-containing genomic DNA clones. In all of these examples the main idea behind the identification of closely linked SNP markers was to enhance the efficiency and cost effectiveness through MAS and increase the resolution within the target locus.

Examples in other Crops

In a study conducted in canola to map the fad2 and fad3 gene, single nucleotide mutations were identified by sequencing the genomic clones of these genes and subsequently SNP markers were developed. Allele-specific PCR assays were developed to enable direct selection of desirable fad2 and fad3 alleles in markerassisted trait introgression and breeding. In barley, SNP markers were identified that were linked to a covered smut resistance gene, Ruh.7H, by using high-resolution melting (HRM) technique. In sugar beet, an anchored linkage map based on AFLP, SNP, and RAPD markers was developed to map QTL for Beet necrotic yellow vein virus resistance genes, Rz4 and Rz5. A consensus genetic map based on EST-derived SNPs was developed for cowpea that would be an important resource for genomic and QTL mapping studies in this crop. In one of the postgenomic era studies in 2002, the fine mapping and mapbased cloning approaches were used to clone the VTC2 gene in Arabidopsis. The authors fine mapped the gene interval from ~980 kb region to a 20 kb interval with SNP and InDel markers. Additional nine candidate genes were identified in that interval and subsequently the underlying mutation was discovered. Although only a few examples that demonstrate the application of SNP markers in QTL mapping and genomic studies have been mentioned here, several other studies have been published in this area. Recent advances in HTP genotyping technologies and sequence information will further pave the way for rapid identification of causative variations and cloning of QTL of interest for use in MAB.

Genome-wide Association Study Approach

GWAS is increasingly becoming a popular tool for dissecting complex traits in plants. The idea behind GWAS is to genotype a large number of markers distributed across the genome so that the phenotype or the functional alleles will be in LD with one or few markers that could then be used in the breeding program. However, due to limited extent of LD, a greater number of markers are required for sufficient power to detect linkage between the marker and the underlying phenotypic variation. Several studies on association mapping in plants have been published and reviewed in the past.

The successful use and first time demonstration of the power of GWAS was through the identification of a putative gene associated with a QTL in maize. In that study, a single locus with major effect on oleic acid was mapped to a 4 cM genetic interval by using SNP haplotypes at 8,590 loci. A fatty acid desaturase gene, fad2, at ~2 kb from one of the associated markers, and this was considered a likely causative gene. With the discovery of millions of SNPs in maize and the availability of tools such as NAM populations, GWAS was effectively applied to dissect the genetic architecture of leaf traits and it was also shown that variations at the liguleless genes contributed to more upright leaf phenotype. Utility of the GWAS approach was demonstrated in barley through the mapping of a QTL for spot blotch disease resistance. Using the diversity array technology (DArT) and SNP markers, several QTL, some of which were not identified for this trait earlier. Another variant of the association mapping method is the CGA, where the association between one or few gene candidate loci and the trait of interest is tested. Using this approach 24 gene candidates were analyzed for association with the field resistance to late blight disease in potato and plant maturity. Nine SNPs were identified to be associated with maturity corrected resistance, explaining 50% of the genetic variance of this trait. Two SNPs at the allene oxide synthase 2 (StAOS2) gene locus were associated with the largest effect on the trait of interest. A GWAS approach was also successfully applied to understand the genetic architecture of complex diseases such as northern and southern corn leaf blights. As GWAS requires large number of molecular markers, the utility of GWAS in dissection of molecular basis of traits in polyploid crops such as canola, wheat, and cotton has been fairly limited due to the insufficient number of polymorphic markers and the absence of reference genome. However, recently developed associative transcriptomics method has a potential to overcome the shortages. Harper leveraged differentially expressed transcriptome sequences to develop molecular markers in tetraploid crop Brassica napus and associated them with glucosinolate content variation in seeds. Due to the precision of this method, scientists were able to correlate specific deletions in canola genome with two QTL controlling the trait. Annotation of deleted regions revealed the orthologs of the transcription factor HAG1, which controlled aliphatic glucosinolate biosynthesis in A. thaliana.

Microsatellite

A microsatellite is a tract of repetitive DNA in which certain DNA motifs (ranging in length from one to six or more base pairs) are repeated, typically 5–50 times. Microsatellites occur at thousands of locations within an organism's genome. They have a higher mutation rate than other areas of DNA leading to high genetic diversity. Microsatellites are often referred to as short tandem repeats (STRs) by forensic geneticists and in genetic genealogy, or as simple sequence repeats (SSRs) by plant geneticists.

Microsatellites and their longer cousins, the minisatellites, together are classified as VNTR (variable number of tandem repeats) DNA. The name "satellite" DNA refers to the early observation that centrifugation of genomic DNA in a test tube separates a prominent layer of bulk DNA from accompanying "satellite" layers of repetitive DNA.

They are widely used for DNA profiling in cancer diagnosis, in kinship analysis (especially paternity testing) and in forensic identification. They are also used in genetic linkage analysis to locate a gene or a mutation responsible for a given trait or disease. Microsatellites are also used in population genetics to measure levels of relatedness between subspecies, groups and individuals.

Structures, Locations and Functions

A microsatellite is a tract of tandemly repeated (i.e. adjacent) DNA motifs that range in length from one to six or up to ten nucleotides (the exact definition and delineation to the longer minisatellites varies from author to author), and are typically repeated 5–50 times. For example, the sequence TATATATATA is a dinucleotide microsatellite, and GTCGTCGTCGTCGTC is a trinucleotide microsatellite (with A being Adenine, G Guanine, C Cytosine, and T Thymine). Repeat units of four and five nucleotides are referred to as tetra- and pentanucleotide motifs, respectively. Most eukaryotes have microsatellites, with the notable exception of some yeast species. Microsatellites are distributed throughout the genome. The human genome for example contains 50,000–100,000 dinucleotide microsatellites, and lesser numbers of tri-, tetra- and pentanucleotide microsatellites. Many are located in non-coding parts of the human genome and therefore do not produce proteins, but they can also be located in regulatory regions and coding regions.

Microsatellites in non-coding regions may not have any specific function, and therefore might not be selected against; this allows them to accumulate mutations unhindered over the generations and gives rise to variability that can be used for DNA fingerprinting and identification purposes. Other microsatellites are located in regulatory flanking or intronic regions of genes, or directly in codons of genes – microsatellite mutations in such cases can lead to phenotypic changes and diseases, notably in triplet expansion diseases such as fragile X syndrome and Huntington's disease.

The telomeres at the ends of the chromosomes, thought to be involved in ageing/senescence, consist of repetitive DNA, with the hexanucleotide repeat motif TTAGGG in vertebrates. They are thus classified as minisatellites. Similarly, insects have shorter repeat motifs in their telomeres that could arguably be considered microsatellites.

Mutation Mechanisms and Mutation Rates

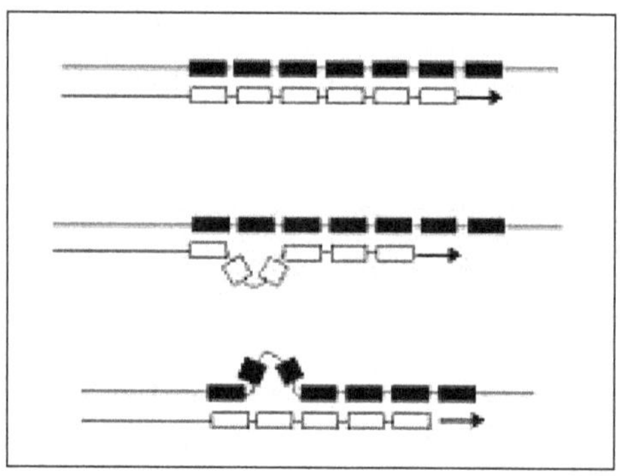

DNA strand slippage during replication of an STR locus. Boxes symbolize repetitive DNA units. Arrows indicate the direction in which a new DNA strand (white boxes) is being replicated from the template strand (black boxes). Three situations during DNA replication are depicted. (a) Replication of the STR locus has proceeded without a mutation. (b) Replication of the STR locus has led to a gain of one unit owing to a loop in the new strand; the aberrant loop is stabilized by flanking units complementary to the opposite strand. (c) Replication of the STR locus has led to a loss of one unit owing to a loop in the template strand.

Unlike point mutations, which affect only a single nucleotide, microsatellite mutations lead to the gain or loss of an entire repeat unit, and sometimes two or more repeats simultaneously. Thus, the mutation rate at microsatellite loci is expected to differ from other mutation rates, such as base substitution rates. The actual cause of mutations in microsatellites is debated.

One proposed cause of such length changes is replication slippage, caused by mismatches between DNA strands while being replicated during meiosis. DNA polymerase, the enzyme responsible for reading DNA during replication, can slip while moving along the template strand and continue at the wrong nucleotide. DNA polymerase slippage is more likely to occur when a repetitive sequence (such as CGCGCG) is replicated. Because microsatellites consist of such repetitive sequences, DNA polymerase may make errors at a higher rate in these sequence regions. Several studies have found evidence that slippage is the cause of microsatellite mutations. Typically, slippage in each microsatellite occurs about once per 1,000 generations. Thus, slippage changes in repetitive DNA are three orders of magnitude more common than point mutations in other parts of the genome. Most slippage results in a change of just one repeat unit, and slippage rates vary for different allele lengths and repeat unit sizes, and within different species. If there is a large size difference between individual alleles, then there may be increased instability during recombination at meiosis.

Another possible cause of microsatellite mutations are point mutations, where only one nucleotide is incorrectly copied during replication. A study comparing human and primate genomes found that most changes in repeat number in short microsatellites appear due to point mutations rather than slippage.

Microsatellite Mutation Rates

Microsatellite mutation rates vary with base position relative to the microsatellite, repeat type, and base identity. Mutation rate rises specifically with repeat number, peaking around six to eight repeats and then decreasing again. Increased heterozygosity in a population will also increase microsatellite mutation rates, especially when there is a large length difference between alleles. This is likely due to homologous chromosomes with arms of unequal lengths causing instability during meiosis.

Direct estimates of microsatellite mutation rates have been made in numerous organisms, from insects to humans. In the desert locust Schistocerca gregaria, the microsatellite mutation rate was estimated at 2.1×10^{-4} per generation per locus. The microsatellite mutation rate in human male germ lines is five to six times higher than in female germ lines and ranges from 0 to 7×10^{-3} per locus per gamete per generation. In the nematode Pristionchus pacificus, the estimated microsatellite mutation rate ranges from 8.9×10^{-5} to 7.5×10^{-4} per locus per generation.

Biological Effects of Microsatellite Mutations

Many microsatellites are located in non-coding DNA and are biologically silent. Others are located in regulatory or even coding DNA – microsatellite mutations in such cases can lead to phenotypic changes and diseases. A genome-wide study estimates that microsatellite variation contributes 10–15% of heritable gene expression variation in humans.

Effects on Proteins

In mammals, 20% to 40% of proteins contain repeating sequences of amino acids encoded by short sequence repeats. Most of the short sequence repeats within protein-coding portions of the genome have a repeating unit of three nucleotides, since that length will not cause frame-shifts when mutating. Each trinucleotide repeating sequence is transcribed into a repeating series of the same amino acid. In yeasts, the most common repeated amino acids are glutamine, glutamic acid, asparagine, aspartic acid and serine.

Mutations in these repeating segments can affect the physical and chemical properties of proteins, with the potential for producing gradual and predictable changes in protein action. For example, length changes in tandemly repeating regions in the Runx2 gene lead to differences in facial length in domesticated dogs (Canis familiaris), with an association between longer sequence lengths and longer faces. This association also applies to a wider range of Carnivora species. Length changes in polyalanine tracts within the HoxA13 gene are linked to Hand-Foot-Genital Syndrome, a developmental disorder in humans. Length changes in other triplet repeats are linked to more than 40 neurological diseases in humans, notably triplet expansion diseases such as fragile X syndrome and Huntington's disease. Evolutionary changes from replication slippage also occur in simpler organisms. For example, microsatellite length changes are common within surface membrane proteins in yeast, providing rapid evolution in cell properties. Specifically, length changes in the FLO1 gene control the level of adhesion to substrates. Short sequence repeats also provide rapid evolutionary change to surface proteins in pathenogenic bacteria; this may allow them to keep up with immunological changes in their hosts. Length changes in short sequence repeats in a fungus (Neurospora crassa) control the duration of its circadian clock cycles.

Effects on Gene Regulation

Length changes of microsatellites within promoters and other cis-regulatory regions can change gene expression quickly, between generations. The human genome contains many (>16,000) short sequence repeats in regulatory regions, which provide 'tuning knobs' on the expression of many genes.

Length changes in bacterial SSRs can affect fimbriae formation in Haemophilus influenzae, by altering promoter spacing. Dinucleotide microsatellites are linked to abundant variation in cis-regulatory control regions in the human genome. Microsatellites in control regions of the Vasopressin 1a receptor gene in voles influence their social behavior, and level of monogamy.

In Ewing's sarcoma (a type of painful bone cancer in young humans), a point mutation has created an extended GGAA microsatellite which binds a transcription factor, which in turn activates the EGR2 gene which drives the cancer. In addition, other GGAA microsatellites may influence the expression of genes that contribute to the clinical outcome of Ewing sarcoma patients.

Effects within Introns

Microsatellites within introns also influence phenotype, through means that are not currently understood. For example, a GAA triplet expansion in the first intron of the X25 gene appears to interfere with transcription, and causes Friedreich Ataxia. Tandem repeats in the first intron of the Asparagine synthetase gene are linked to acute lymphoblastic leukaemia. A repeat polymorphism in the fourth intron of the NOS3 gene is linked to hypertension in a Tunisian population. Reduced repeat lengths in the EGFR gene are linked with osteosarcomas.

An archaic form of splicing preserved in Zebrafish is known to use microsatellite sequences within intronic mRNA for the removal of introns in the absence of U2AF2 and other splicing machinery. It is theorized that these sequences form highly stable cloverleaf configurations that bring the 3' and 5' intron splice sites into close proximity, effectively replacing the spliceosome. This method of RNA splicing is believed to have diverged from human evolution at the formation of tetrapods and to represent an artifact of an RNA world.

Effects within Transposons

Almost 50% of the human genome is contained in various types of transposable elements (also called transposons, or 'jumping genes'), and many of them contain repetitive DNA. It is probable that short sequence repeats in those locations are also involved in the regulation of gene expression.

Analysis

Repetitive DNA is not easily analysed by next generation DNA sequencing methods, which struggle with homopolymeric tracts. Therefore, microsatellites are normally analysed by conventional PCR amplification and amplicon size determination, sometimes followed by Sanger DNA sequencing.

In forensics, the analysis is performed by extracting nuclear DNA from the cells of a sample of interest, then amplifying specific polymorphic regions of the extracted DNA by means of the polymerase chain reaction. Once these sequences have been amplified, they are resolved either through gel electrophoresis or capillary electrophoresis, which will allow the analyst to determine how many repeats of the microsatellites sequence in question there are. If the DNA was resolved by gel electrophoresis, the DNA can be visualized either by silver staining (low sensitivity, safe, inexpensive), or an intercalating dye such as ethidium bromide (fairly sensitive, moderate health risks, inexpensive), or as most modern forensics labs use, fluorescent dyes (highly sensitive, safe, expensive).

Instruments built to resolve microsatellite fragments by capillary electrophoresis also use fluorescent dyes. Forensic profiles are stored in major databanks. The British data base for microsatellite loci identification was originally based on the British SGM+ system using 10 loci and a sex marker. The Americans increased this number to 13 loci. The Australian database is called the NCIDD, and since 2013 it has been using 18 core markers for DNA profiling.

Amplification

Microsatellites can be amplified for identification by the polymerase chain reaction (PCR) process, using the unique sequences of flanking regions as primers. DNA is repeatedly denatured at a high temperature to separate the double strand, then cooled to allow annealing of primers and the extension of nucleotide sequences through the microsatellite. This process results in production of enough DNA to be visible on agarose or polyacrylamide gels; only small amounts of DNA are needed for amplification because in this way thermocycling creates an exponential increase in the replicated segment. With the abundance of PCR technology, primers that flank microsatellite loci are simple and quick to use, but the development of correctly functioning primers is often a tedious and costly process.

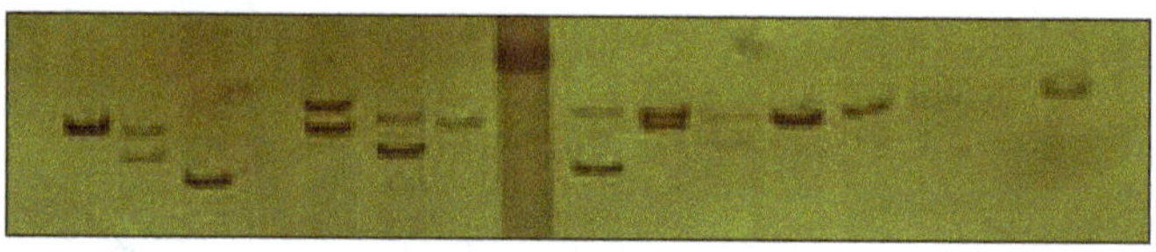

A number of DNA samples from specimens of Littorina plena amplified using polymerase chain reaction with primers targeting a variable simple sequence repeat (SSR, a.k.a. microsatellite) locus. Samples were run on a 5% polyacrylamide gel and visualized using silver staining.

Design of Microsatellite Primers

If searching for microsatellite markers in specific regions of a genome, for example within a particular intron, primers can be designed manually. This involves searching the genomic DNA sequence for microsatellite repeats, which can be done by eye or by using automated tools such as repeat masker. Once the potentially useful microsatellites are determined, the flanking sequences can be used to design oligonucleotide primers which will amplify the specific microsatellite repeat in a PCR reaction.

Random microsatellite primers can be developed by cloning random segments of DNA from the focal species. These random segments are inserted into a plasmid or bacteriophage vector, which is in turn implanted into Escherichia coli bacteria. Colonies are then developed, and screened with fluorescently–labelled oligonucleotide sequences that will hybridize to a microsatellite repeat, if present on the DNA segment. If positive clones can be obtained from this procedure, the DNA is sequenced and PCR primers are chosen from sequences flanking such regions to determine a specific locus. This process involves significant trial and error on the part of researchers, as microsatellite repeat sequences must be predicted and primers that are randomly isolated may not display significant polymorphism. Microsatellite loci are widely distributed throughout the genome and can be isolated from semi-degraded DNA of older specimens, as all that is needed is a suitable substrate for amplification through PCR.

More recent techniques involve using oligonucleotide sequences consisting of repeats complementary to repeats in the microsatellite to "enrich" the DNA extracted (Microsatellite enrichment). The oligonucleotide probe hybridizes with the repeat in the microsatellite, and the probe/microsatellite complex is then pulled out of solution. The enriched DNA is then cloned as normal, but the proportion of successes will now be much higher, drastically reducing the time required to develop the regions for use. However, which probes to use can be a trial and error process in itself.

ISSR-PCR

ISSR (for inter-simple sequence repeat) is a general term for a genome region between microsatellite loci. The complementary sequences to two neighboring microsatellites are used as PCR primers; the variable region between them gets amplified. The limited length of amplification cycles during PCR prevents excessive replication of overly long contiguous DNA sequences, so the result will be a mix of a variety of amplified DNA strands which are generally short but vary much in length.

Sequences amplified by ISSR-PCR can be used for DNA fingerprinting. Since an ISSR may be a conserved or nonconserved region, this technique is not useful for distinguishing individuals, but rather for phylogeography analyses or maybe delimiting species; sequence diversity is lower than in SSR-PCR, but still higher than in actual gene sequences. In addition, microsatellite sequencing and ISSR sequencing are mutually assisting, as one produces primers for the other.

Limitations

Repetitive DNA is not easily analysed by next generation DNA sequencing methods, which struggle with homopolymeric tracts. Therefore, microsatellites are normally analysed by conventional PCR amplification and amplicon size determination. The use of PCR means that microsatellite length analysis is prone to PCR limitations like any other PCR-amplified DNA locus. A particular concern is the occurrence of 'null alleles':

- Occasionally, within a sample of individuals such as in paternity testing casework, a mutation in the DNA flanking the microsatellite can prevent the PCR primer from binding and producing an amplicon (creating a "null allele" in a gel assay), thus only one allele is amplified (from the non-mutated sister chromosome), and the individual may then falsely appear to be homozygous. This can cause confusion in paternity casework. It may then be necessary to amplify the microsatellite using a different set of primers. Null alleles are caused especially by mutations at the 3' section, where extension commences.

- In species or population analysis, for example in conservation work, PCR primers which amplify microsatellites in one individual or species can work in other species. However, the risk of applying PCR primers across different species is that null alleles become likely, whenever sequence divergence is too great for the primers to bind. The species may then artificially appear to have a reduced diversity. Null alleles in this case can sometimes be indicated by an excessive frequency of homozygotes causing deviations from Hardy-Weinberg equilibrium expectations.

Development and Application of Microsatellites in Plant Breeding

Hybridization method is very commonly used in crop breeding in order to obtain segregant populations portraying interesting agronomical traits and broad genetic variability. This method combines available genes from two or more different genitors in a unique individual.

The choice of genitors is very important in this process because it assures the obtaining of promising segregant populations. This procedures is based on genetic dissimilarity of the subject under study. In plants, genetic diversity may be inferred through quantitative (phenotypic ou genotypic traits) and predictive charecteristics. The quantitative genetic diversity is usually calculated from a set of genotypic or phenotypic characters, throught the use multivariate statistics. Predictive methods are those based on morphological, physiological or molecular differences, quantifying then in some similarity/ dissimilarity measurement, which may express the degree of parental genetic diversity.

However, due to high environmental influence on morphoagronomic characteristics, many researchers have chosen predictive techniques for estimating genetic diversity as they reduce the necessity of hybridization among progenitors, diminishing labor and expenses with a great number of crosses and field experiments.

Molecular markers are powerful tools for analyzing genome diversity within a species, and to evaluate genetic relationships between individuals and populations. They have been used for mapping of genomic regions containing genes of agricultural interest. In recent years, genetic divergence studies in different crops have often used microsatellite as the main tool for genetic diversity analysis.

Microsatellites or SSRs are one of the most indicated techniques to study polymorphism between DNA sequences. These molecular markers are based on PCR reaction that detects locivariations of repetitive sequences. They present high levels of polymorphism, codominant inheritance, multiallelism, mendelian pattern and good genome coverage. Microsatellites require low amount of DNA, can be easily automated for high throughput screening, may be exchanged between laboratories, and are highly transferable between populations.

SSRs occur frequently within or near genes in plants. Comparing coding and non-coding regions in different plant species, it was observed that tri and tetra-nucleotide microsatellite motives are more common within introns, whereas other types of motives are found within exons. Morgante reported a selective pressure acting over untranslated 3' and 5' regions (UTR) of genes, favoring a higher pressure of SSRs in these regions than in the rest of the genome. The 5'-UTR presents a higher CT and CTT number of repetitions, while the 3'UTR showed to be rich in AG and AAG. Regions of DNA involved in gene regulation are expected to exhibit sequence conservation between related species over evolutionary time due to functional constraints. Microsatellite transferability amongst related species is allowed by the homologous nature of the DNA sequence in microsatellite flanking regions.

Microsatellite Enriched Libraries Development

The development of microsatellite markers involves several distinct steps from obtaining the library to developing a working set of primers that can amplify polymorphic microsatellite loci.

Traditional methods for developing microsatellites have a low efficiency and can be time-consuming, especially in the selection stage and clone screening. Ostrander proposed the enrichment methodology for microsatellites achieving enrichment 50 times superior to genomic libraries. The selective hybridization protocols are very commonly used in microsatellite enrichment.

Construction of an Enriched Microsatellite Library for Common Beans

An enriched library was constructed according to Billotte for IAC-UNA according to Benchimol. The first step was to quantify the genomic DNA (200 ng mL) extracted using CTAB method on agarose gel. The selection of microsatellite sequences was performed using biotin-labeled microsatellite oligoprobe and streptavidin-coated magnetic beads following a hybridization-based capture methodology.

Plasmids form positive clones containing microsatellites were extracted by plasmid DNA minipreparation. All clones were sequenced in both directions. Reads were processed by the Phred version 0.000925.c base calling program and vector sequences, poly-A tail, and adapters were trimmed after cross-match analysis. Only perfect and imperfect sequences containing to five or more repeated units were considered. Complementary primers to single sequences, flanking the microsatellites, were designed. Stringency criteria adopted was based on GC content between 40 and 60 %, melting temperature between 46 and 60 °C, a salt concentration of 50 mM, and a product length between 150 and 300 pb.

Genic Microsatellites from EST Database

The major constraint of using SSR markers from genomic libraries is the high development cost and effort required to obtain working primers for a given study species. With the establishment of expressed sequence tag (EST) sequence projects for many plant species, a great amount of information and DNA sequences were generated and deposited in online databases. The development of EST-based SSR markers through data mining has become a fast, efficient, and relatively inexpensive compared with the development of genomic SSRs. Genbank, for instance, gathers a great amount of EST sequences, gene and cDNA clones. The TIGR Institute possesses an EST database for many cultivated plant species of economic importance and therefore, it is hard to find sequences of wild species.

Many programs that have been developed for recognition of SSR patterns in the sequence files, such as the 'Tandem Repeats Finder' - TRF, the 'Simple Sequence Repeat Identification Tool' - SSRIT, and program MISA (Micro Satellite) identification tool. A major disadvantage of EST-derived microsatellites is the sequence redundancy that yields multiple set of markers at the same locus. Therefore it is very important to verify the existence of redundancy among sequences and BLASTN tool could be used for that purpose.

Application of Microsatellites in Plant Breeding

Microsatellites have been used for plant genetic analysis such as to measure the effects of natural selection, to unveil the genetic diversity, to measure population structure, to integrate the genetic, physical and sequence-based physical maps and for marker assisted selection.

Using Microsatellite in Genetic Diversity and Population Structure in Common Bean

Molecular marker analysis has contributed for the understanding of common bean genetic structure diversity and phylogenetic analysis. For molecular markers, such as RAPD, SSR, and RFLP, there were major findings such as plant population continuous variation and high level of dissimilarity within population and among germplasm collections. Microsatellite markers are involved not only in genetic diversity studies, population genetics and evolutionary studies, but are also being used in fundamental research like genome analysis, gene mapping and marker-assisted selection.

Given the high polymorphism level of microsatellites when compared to other markers, many worldwide researchers have preferred using these markers on the study of common bean population structure, genetic divergence and genetic mapping.

Genetic Divergence and Population Structure in Common Bean using Microsatellite

The assessment of genetic diversity and population structure of 122 accessions of common bean from Brazil and Africa was performed using microsatellites, genetic distance estimates and probabilistic models to analyze population structure.

Sixteen microsatellite markers from all the 11 linkage groups of P. vulgaris L. gene map were selected for analyses based on their dispersed map location. Markers originated from genic and genomic sequences were chosen in equal proportions. Microsatellite analysis was conducted, including an economic method of microsatellite fluorescent labeling of fragments amplified by PCR.

Microsatellite methods employing fluorescent labeling and automated band calling with precise software-based allele detection are considered the most accurate way of genotyping. Additionally, this sort of labeling organizes markers into distinct dye color panels allowing multiplexing during band separation, with advantages of high throughput genotyping and simultaneous analysis of multiple loci.

The methodology of multiplexing can also be carried out in PCR amplification step with mixtures of the appropriate primers. Current technology allows capillary-based separation of microsatellite bands with four different color panels, one for each individual marker, which can be evaluated through a single capillary with a separate size standard, allowing precise band size estimates.

The amplified fragments were multiplexed depending on their size variation and analyzed in an ABI 3730 (Applied Biosystems). Genotypes of markers were determined using the GeneMarker program (version1.51, SoftGenetics). The genetic relationship among entire accessions was analyzed by principal coordinate analysis performed using the GenAlex 6 Program.

The genetic distance among 122 Andean and Mesoamerican accessions was calculated using the C.S. Chord distance. The C.S. Chord distance does not require any mutation model for microsatellite evolution and is free from bottleneck effect. Based on this genetic distance, an unrooted neighbor-joining (NJ) tree was constructed in PowerMarker. Distinct clusters were apparent with the SSR markers unambiguously assigning accessions to the Andean and Mesoamerican gene pools with the 2D plot of the PCoA based on pairwise genetic distances.

The obtained results demonstrated that Andean gene pool accessions from Brazil and Africa were clustered together showing a great genetic similarity. Likewise, the Mesoamerican accessions from both countries formed a second group. These results indicate the possibility of introgression between countries. Some African accessions with low genetic similarity could be use in breeding programs to broaden the genetic base of Brazilian cultivars.

Potential Application of SSR Markers for Mapping and Tagging Disease Resistance Genes Common Bean

Anthracnose (ANT) caused by Colletotrichum lindemuthianum (Sacc.) Scrib and angular leaf spot (ALS) caused by Pseudocercospora griseola (Sacc.) Crous e U. Braun, are the most widespread, recurrent and devastating diseases of common bean in Latin America and Africa. Use of disease resistance genes is the most practical, cost-effective and environmentally friendly strategy for the control of ANT and ALS. Resistance to various diseases in common bean is conferred mostly by single, dominant genes with race-specific resistance (R-genes). Resistance to C. lindemuthianum is conditioned by some 13 reported genes identified by the Co symbol. Six independent dominant genes identified by the Phg symbol condition resistance to P. griseola.

Recent studies were conducted to study the co-segregation between ANT and ALS resistance genes that confer resistance to C. lindemuthianum and P. griseola, respectively. A total of 112 individual F_2 derived from a cross between resistant and susceptible parents were inoculated with races 73 of C. lindemuthianum and 63-39 of P. griseola. The analysis was performed using 22 SSR markers. Two contrasting DNA bulks were constructed by pooling equal volumes of fluorometrically standardized DNA from four to six resistant and susceptible individuals, respectively, of the F_2 population. Of these 22 SSR markers, the g2303 showed contrasting amplification patterns in parental materials and in resistant vs. susceptible bulks or individuals. Segregation analyses of the disease reaction from F_2 plants were performed by the chi-square (X^2) test, according to a Mendelian segregation hypothesis of 3 R_ (resistant) to 1 rr (susceptible). Linkage analyses were performed using the computer software Mapmaker/exp 3.0. The results of molecular mapping revealed that the SSR marker g203 was linked at 0.9 cM from ANT resistance gene and 1.8 cM from ALS resistance gene in coupling phase with both genes, and mapped at linkage group Pv04 in the common bean linkage map. The co-segregation between these genes revealed the possibility of monitoring those genes indirectly through the use of g2303 molecular marker. Similar results were obtained by Gonçalves-Vidigal et al.. In this study, the findings might help plant breeding programs to reduce time and cost associated with pyramiding resistance genes to different pathogens.

CRISPR-Cas9

The CRISPR-Cas9 system is a plant breeding innovation that uses site-directed nucleases to target and modify DNA with great accuracy. Developed in 2012 by scientists from the University of California, Berkeley, CRISPR-Cas9 has received a lot of attention in recent years due to its range of applications, including biological research, breeding and development of agricultural crops and animals, and human health applications. These include gene silencing, DNA-free CRISPR-Cas9 gene editing, homology-directed repair (HDR), and transient gene silencing or transcriptional repression (CRISPRi).

CRISPR, or Clustered Regularly Interspaced Short Palindromic Repeats, is an integral part of a bacterial defense system. It is also the basis of the CRISPR-Cas9 system.

The CRISPR molecule is made up of short palindromic DNA sequences that are repeated along the molecule and are regularly-spaced. Between these sequences are "spacers", foreign DNA sequences from organisms that have previously attacked the bacteria. The CRISPR molecule also includes CRISPR-associated genes, or Cas genes. These encode proteins that unwind DNA, and cut DNA, called helicases and nucleases, respectively. The CRISPR immune system protects the bacteria from repeated virus attacks thru three steps:

- Adaptation: When DNA from a virus invades the bacteria, the viral DNA is processed into short segments and is made into a new spacer between the repeats. These will serve as genetic memory of previous infections.
- Production of CRISPR RNA: The CRISPR sequence undergoes transcription, including spacers and Cas genes, creating a single-stranded RNA. The resulting single-stranded RNA is called CRISPR RNA, which contains copies of the invading viral DNA sequence in its spacers.
- Targeting: The CRISPR RNAs will identify viral DNA and guide the CRISPR-associated proteins to them. The protein then cleaves and destroys the targeted viral material.

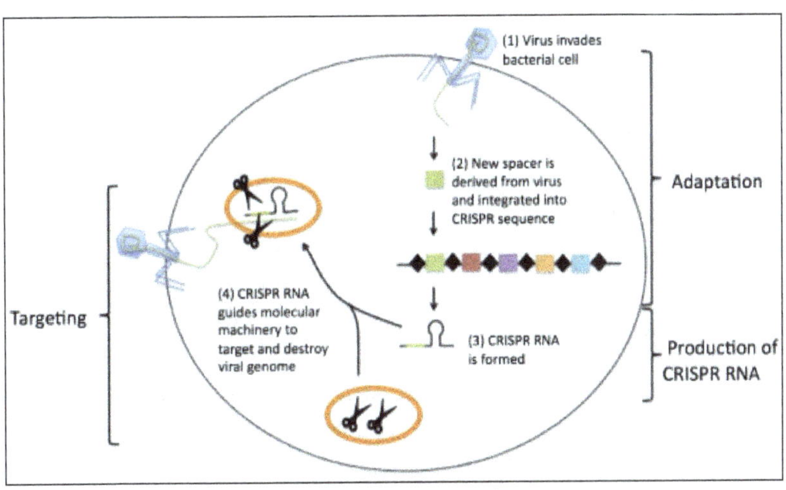

The steps of CRISPR-mediated immunity.

Scientists make use of the CRISPR-Cas9 systems' recognition of specific DNA sequences and apply it in the process of development of improved crops. Instead of viral DNA as spacers, scientists design their own sequences, based on their specific gene of interest. If a gene's sequence known, it can be easily used in CRISPR. It will then act just like a spacer for the system and guide the Cas9 protein to a DNA matching sequence.

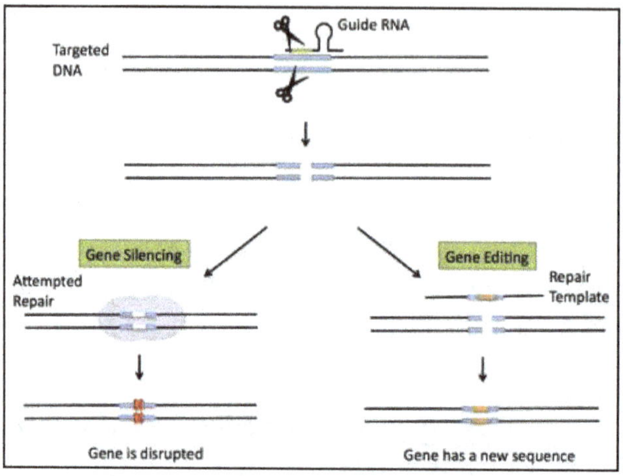

Mechanism of Gene Editing of CRISPR/Cas96.

CRISPR-Cas9 allows researchers to perform the following:

Gene Knock-out

Gene silencing using CRISPR starts with the use of a single guide RNA (sgRNA) to target genes and initiate a double stranded break using the Cas9 endonuclease. These breaks are then repaired by an innate DNA repair mechanisms, the non-homologous end-joining (NHEJ). However, NHEJ is error-prone and results in genomic deletions or insertions, which then translates into permanent silencing of the target gene.

DNA-free Gene Editing

CRISPR can be used for DNA-free gene editing without the use of DNA vectors, requiring only RNA or protein components. A DNA-free gene editing system can be a good choice to avoid the possibility of undesirable genetic alterations due to the plasmid DNA integrating at the cut site or random vector integrations.

Gene Insertions or Knock-ins

The CRISPR-induced double-strand break can also be used to create a gene "knock-ins" by exploiting the cells' homology-directed repair. The precise insertion of a donor template can alter the coding region of a gene.

Transient Gene Silencing

By modifying the Cas9 protein so it cannot cut DNA, transient gene silencing or transcriptional repression can also be done. The modified Cas9, led by a guide RNA, targets the promoter region of a gene and reduces transcriptional activity and gene expression. Transient activation or upregulation of specific genes can be effectively done.

CRISPR-Cas9 Applications

Researchers have found that the CRISPR-Cas9 system can be applied to nearly every organism. Early studies using CRISPR-Cas9 for gene editing have focused on crops important for agriculture. It was realized early on that the system could be used in crops to improve traits, such as yield, plant architecture, plant aesthetics, and disease tolerance.

CRISPR has been used to edit the genome of rice. The team of Ying Wang from Syngenta Biotechnology China designed several CRISPR sgRNAs and successfully deleted fragments of the dense and erect panicle1 (DEP1) gene in the Indica rice line IR58025B. Improvements in yield-related traits, such as dense and erect panicles and reduced plant height,were observed in the mutant plants produced.

A team of researchers from the Chinese Academy of Agricultural Sciences led by Yupeng Cai also used the CRISPR-Cas9 system to induce mutations on GmFT2a, an integrator in the photoperiod flowering pathway of soybean. The developed soybean plants showed late flowering, resulting in increased vegetative size. The mutation was also found to be stably inherited in the following generation.

Researchers from Beijing Key Laboratory of Vegetable Germplasm Improvement, led by Shouwei Tian used CRISPR-Cas9 to target ClPDS, the phytoene desaturase in watermelon, to achieve the albino phenotype. All genome-edited watermelons harbored mutations in ClPDS and showed full or mosaic albino phenotype. This study served as a proof of concept of using the CRISPR-Cas9 system in watermelon breeding.

Researchers from the Chinese Academy of Agricultural Sciences and National Center for Citrus Variety Improvement and Southwest University have also developed citrus plants resistant to citrus canker caused by Xanthomonas citri subsp. citri (Xcc), a serious disease of citrus, through CRISPR-Cas9.

The team targeted the promoter of the CsLOB1 gene, which promotes canker development in citrus. The developed lines showed enhanced resistance to citrus canker compared to wild types.

Cold Spring Harbor Laboratory, together with various research institutions, also used CRISPR-Cas9 to generate mutations in the flowering suppressor SELF-PRUNING5G (SP5G) in tomato to manipulate photoperiod response. The mutations brought about by CRISPR-Cas9 caused rapid flowering and enhanced the compact growth habit of field tomatoes, resulting in a quick burst of flower production and early yield.

CRISPR-Cas9 has also allowed generation of animals suitable for human disease modeling. The team of Yuyu Niu from Yunnan Key Laboratory of Primate Biomedical Research applied CRISPR-Cas9 via coinjection of Cas9 mRNA and sgRNAs into one-cell-stage embryos. The team generated CRISPR-edited cynomolgus monkeys for brain disorders that cannot be fully studied in mice.

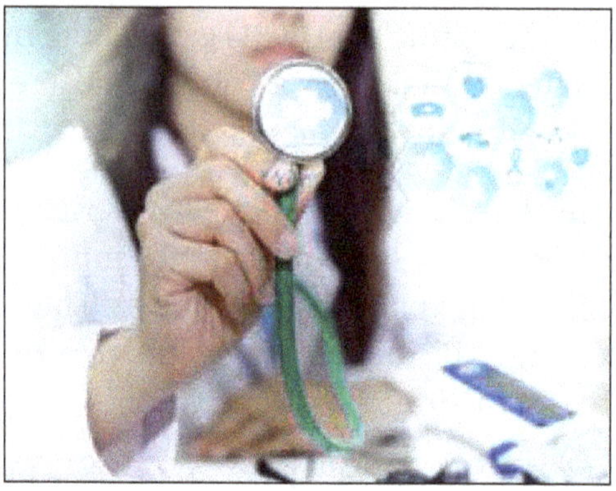

Researchers from the University of California also used CRISPR for studies in gene therapy. Using CRISPR-Cas9, they corrected mutations associated with the genetic disease, β-thalassemia by creating induced pluripotent stem cells (iPSCs) from the β-thalassemia patients. The team used CRISPR-Cas9 to correct mutations in the human hemoglobin beta (HBB) in patient iPSCs, resulting in gene-corrected iPSCs with restored expression of the HBB gene, which can be used for gene therapy.

US scientists are also studying the use of CRISPR for treating the Human Immunodeficiency Virus (HIV). They used CRISPR to edit the HIV genome out of immune cells, called T cells, from an HIV patient. Scientists found that CRISPR can prompt the HIV virus to mutate. However, more studies are still needed before CRISPR can be used to treat HIV.

CRISPR has played a huge part in the increase in genome editing studies in recent years. The system has broad applications in plant and animal improvement, as well as in the medical field. As a relatively young technique, various discoveries and innovations for its efficient use in wider applications are in the offing.

Genomic Selection Strategy

Marker-assisted selection (MAS) is an important scheme in plant breeding since the 1990s, after promising analysis results for tagging genes or mapping QTL. Marker assisted selection and molecular breeding have been used in the identification of underlying major genes in gene pools and their transfer to desirables traits of major plant breeding programs. Using of MAS have shown some shortcomings due to long selection schemes and also the look for vital marker-QTL associations being unable to capture "minor" gene effects. Thus marker-assisted selection (MAS) is difficult to improve traits having complex inheritance such as grain yield and abiotic stresses.

Using whole-genome prediction models, the genomic selection (GS) strategy has paved the way to over-come these limitations. High-density molecular markers using is one of the main features of genomic selection. Therefore, each of the trait loci has the likelihood of being in linkage disequilibrium (LD) with a minimum of one marker locus within the entire breeding population. Genome selection strategy removes the need to mapping of genes and search for linked QTL–marker loci associated individually. Rather, Genomic selection accounts for bunches of predictors simultaneously and is characterized by constraining random estimates towards zero. Moreover, Genomic selection helpful for accelerate breeding cycles in such a way that the rate of annual genetic gain per unit of time and cost can be decreased. Genomic selection has been well established in the field of animal breeding, but is in its beginning in crops plants and forest tree breeding.

Genome-wide selection or genomic selection estimates marker effects across the full ordering of the breeding population (BP) supported the prediction model developed within the training population (TP). Training population could be a group of related individuals (such as half-sibs or lines) that are each phenotypes and genotypes. Breeding population typically is just genotyped not phenotypes. Hence, Genomic selection depends on the degree of genetic similarity between training population and breeding population within the Linkage disequilibrium between marker and trait loci. Breeding values have not been a preferred index in plant breeding, however it is in animal breeding. Once plan of genomic estimated breeding value (GEBV) was planned, it had been considered an unrealistic approach due to lack of enormous scale genotyping technologies. However, currently, it has been a possible approach with recent advances in high throughput genotyping platforms (3rd generation platforms). Generally processes of genomic selection and marker assisted selection used for Quantitative Traits are shown in figure.

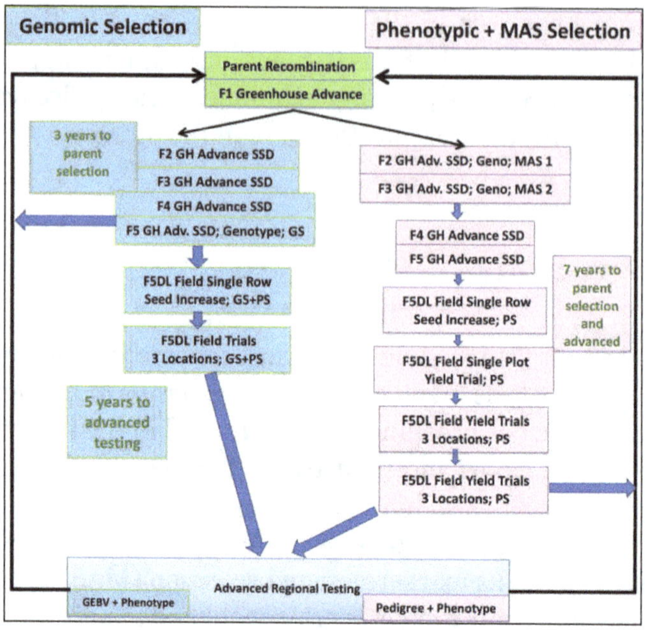

Self-pollinated crop genomic selection vs. phenotypic/MAS selection timeline.

The main schemes of the two approaches are similar, wherever each marker assisted selection and genomic selection consist of breeding and training phases. In the training phase, phenotypes and genome-wide (GW) genotypes are investigated in an exceedingly set of a population, i.e., the mapping population in marker assisted selection and also the training population in genomic selection. Among populations, important relationships between phenotypes and genotypes are expected utilizing statistical models. Within the breeding phase, genotype data are obtained in an exceedingly breeding population, on the basis of genotypic information favorable individuals are selected. There are three prominent variations between the two approaches: (1) within the training section, quantitative trait loci (QTLs) are known in marker assisted selection whereas formulae for genetic estimation of breeding value prediction are generated in genomic selection, called genomic selection models; (2) within the breeding section, genotype data are solely needed for targeted regions in marker assisted selection, whereas genomic selection genotype data are considered to be mandatory in genomic selection (3) within the breeding phase, favorable individuals are selected on the bases of the linked markers in marker assisted selection, whereas GEBVs are used for selection in GS. Thus, GS collectively analyses all the genetic variance of every individual by summing the marker impacts of GEBV and it is expected to deal with little effect genes that cannot be captured by traditional MAS.

The statistical ways employed by GS are comparatively new the plant-breeding community. The ways of marker-assisted selection (MAS) or marker-assisted recurrent selection (MARS) assume that the user is aware of that alleles are favorable, and what their average effects on the phenotype are. This assumption is viable for major-gene traits however not for quantitative traits that are influenced by several loci of little impact and the environment. To deal with quantitative traits, new statistical approaches that might account for this uncertainty were required to get the most effective predictions potential. Finding problem with locus identification, entailed that the consequences for all marker loci be at the same time estimated. Once a prediction based on allele effects, the allele becomes the unit of analysis. Alleles are so the units that need to be replicated inside and across environments. However that replication will occur in spite of the particular lines carrying the alleles such lines themselves no longer need to be replicated. Within the breeding context, removing the requirement for line replication opens the likelihood of dramatically increasing the amount of lines pushed through the pipeline of a breeding program, and successively of accelerating selection intensity.

Genomic Selection Scheme

Genomic selection is to assemble a training population for individuals for which both genotypes and phenotypes are available and use those data to create a statistical model that relates variation in observed genotypes marker loci to variation in the observed phenotypes of the individuals. Multiple generations of parents and progenies provided powerful training population than a single generation individuals and larger number of individual's generations and markers provide more powerful training population (TP). The statistical model obtained from genotype and phenotype is then applied to a prediction population comprised of individuals for which genotypes are available, but phenotypes are not. GS is based on similarity between the training population (TP) and breeding population (BP) in the LD between marker loci and trait loci. This similarity may exist because breeding population is selected from training population or descended from training population or because density of markers is so high that every trait locus is in disequilibrium with at least one marker locus across the entire population of the target species. The training population is genotyped and phenotyped to train the genomic selection (GS) prediction model. In Genomic selection main role of phenotyping is to calculate effect of markers & cross validation. Genotypic information from the breeding material is then fed into the model to calculate genome estimated breeding values (GEBV) for these lines.

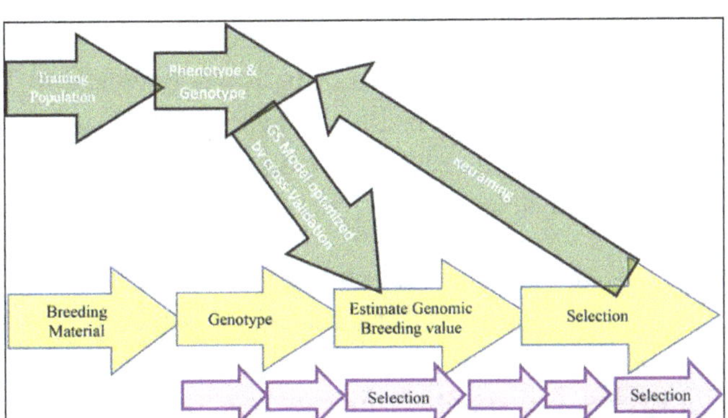

Genomic selection scheme. Information on phenotype and genotype for a training population allows estimating parameters for the model.

Need of Genomic Selection

Traditional marker assisted selection, whereas helpful for merely transmitted traits controlled by few loci, loses effectiveness because the number of loci will increase. This is often true for individual quantitative traits or once multiple traits are below selection. Quantitative traits like grain yield, abiotic stress have verified hard to enhance with marker-assisted selection. The main limitations are (i) tiny population sizes and traditional statistical strategies that have inadequate power to find and accurately estimate effects of small-effect quantitative trait loci (QTL) and (ii) gene x gene interactions (epistasis) and (iii) genotype x environment interactions (G.E) that have restricted the exchangeability of quantitative trait loci result estimates across populations and environments. The Beavis effect is a statistical phenomenon in biology that refers to the overestimation of the effect size of quantitative trait loci (QTL) as a result of small sample sizes in QTL studies.

The availability of low cost and extensive molecular markers in plants has allowed breeders to raise however molecular markers might best be used to win breeding progress. Additionally advances in high-throughput genotyping have markedly reduced the value per data point of molecular markers and increasing genome coverage. This reduction was in the main the results of three parallel developments (i) the invention of huge numbers of single nucleotide polymorphism (SNP) markers in several species; (ii) development of high-throughput technologies, like multiplexing and gel-free deoxyribonucleic acid arrays, for screening SNP polymorphisms; and (iii) automation of the marker-genotyping method, together with efficient procedures for deoxyribonucleic acid extraction. Phenotyping prices are increased Genotyping prices are being reduced and marker densities are being increased speedily.

Statistical strategies are inadequate for improving polygenic traits controlled by several loci of small impact. There will be more markers (explanatory variables) than lines (observations) that introduce statistical issues. Drawback of small p (number of traits) and enormous m (number of markers) ends up in a lack of degrees of freedom. The foremost acceptable statistical model is required to at the same time estimate several marker effects from a limited range of phenotypes. In so-called "large p, small m" problems, standard multiple linear regression cannot be used without variable selection, that conflicts with the initial goal of avoiding marker selection. To overcome these issues, a range of ways, e.g., best linear unbiased prediction, ridge regression, Bayesian regression, kernel regression and machine learning methods are projected to develop prediction models for genomic selection.

The most economical use of GS is to exchange expensive and long phenotyping by a prediction of the genetic worth of the character below selection (or any multi trait index). Thus, the foremost expected advantage is to shorten selection cycles. However, to learn from shorter cycles, the genetic gain per selection cycle ought to be near that predicted from phenotypic or combined MAS + phenotypic selection. Progeny testing schemes have a high accuracy of selection, however the time interval is also additional, takes long term to perform a cycle of selection that decreases the genetic gain. The univariate breeder's equation was used for the GS-BPs as a result of they include just one stage of selection. Selection accuracy is adequate to the correlation between selection criteria and breeding value (i.e., correlation between phenotypes or GEBVs and true breeding values [TBVs]). In oxen, Schaeffer determined that the time and value savings exploitation GS with GEBV accuracy of 0.75 would increase genetic gain twofold and supply a price savings of ninety two in comparison to the present ways. The power to calculate extremely correct GEBVs and also the potential to drastically cut back make-up analysis frequency and selection cycle time expedited a speedy adoption of genomic selection and is revolutionizing the oxen breeding trade.

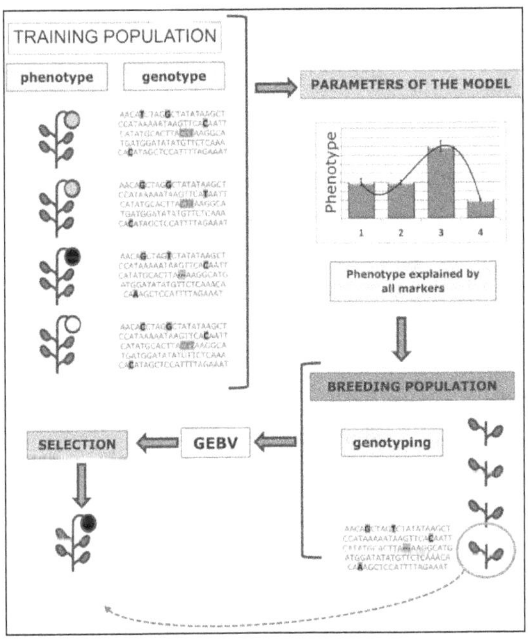

Genomic selection scheme.

Model for Genomic Selection

The basic model may be denoted as,

$$Y_i = g(x_i) + e_i$$

where Y_i is an observed phenotype of individual i (i = 1 ... n) and x_i is a 1 x pvector of SNP genotypes on individual i, $g(x_i)$ is a function relating genotypes to phenotypes, and e_i a residual term. The GEBV is generally equal to g (x_i). Further similarities among GS models can be seen by recognizing that they all seek to minimize a certain cost function. In least squares analysis, the well-known cost function is simply the sum of squared residuals.

Cross-validation

Evaluating GEBV accuracy through cross validation (CV). CV entails splitting the data into training and validation set. The ratio of observations in each set varies, but often a fivefold CV is used, that is, the data set is randomly divided into five sets, with four sets being combined to form the training set and the remaining set designated as the validation set. Each subset of the data is used as the validation set once, before applying of the prediction model to the breeding population, the accuracy of the model should be tested. For this, most of the training population is used to create a prediction model, which is then used to estimate the genomic estimation breeding values of the remaining individuals in the training population, using genotypic data only. This permits researchers to "test" and refine the prediction model to make sure the prediction accuracy is high enough that future predictions are often relied upon. Once valid, the model is often applied to a breeding population to calculate GEBVs of lines that genotypical, however no phenotypical, information is available.

Genomic Selection Prediction Accuracies

The prediction accuracy of the GEBVs is evaluated by the correlation between the GEBVs and empirically estimated breeding values, r(GEBV: EBV), where the EBV can be obtained in a number of ways, most simply, as a phenotypic mean. This correlation provides an estimate of selection accuracy and thus directly relates GEBV prediction accuracy to selection response. Other statistics such as mean-square error (MSE) are used occasionally. Genomic selection accuracy is defined as the correlation between GEBV and the true breeding value (TBV), that is, r(GEBV:TBV). Since we can only measure r(GEBV:EBV), this measure needs to be converted to an estimate of r(GEBV:TBV). To do so, it is assumed that r(GEBV:EBV) = r(GEBV:TBV) X r(EBV:TBV). This assumption is correct if the only component common between the GEBV and the EBV is the TBV itself. In other words, the assumption holds if GEBV = TBV + e1 and EBV = TBV + e2, where e1 and e2 are uncorrelated residuals. The assumption could be violated if the training and validation data were collected in the same environment. In that case, genotype by environment (G.E) interaction would generate a common component of error in both GEBV and EBV, biasing their correlation upward. Thus, training and validation data should be collected in different environments to ensure sound estimates of GEBV prediction accuracy. The correction, r(EBV:TBV), accounts for the fact that the EBV in the validation set is not free of error. When the EBVs are phenotypes, r(EBV:TBV) is equal to the square root of heritability (h) within the validation set.

Accurate GEBV predictions offer the possibility that future elite and parental lines will be selected on GEBV rather than phenoypic data from extensive field testing. Immediate impact would be a great increase in speed of breeding cycle increasing selection gains per unit time. Thus, GS could radically change the practice of field evaluation for breeders. Of course, regardless of the breeding method used, final field evaluations of varieties across the target environments will be needed before they are distributed to farmers.

Breeding cycle time is shortened by removing phenotypic evaluation of lines before selection as parents for the next cycle. Model training and line development cycle length will be crop and breeding program specific. In a GS breeding schema, genome-wide DNA markers are used to predict which individuals in a breeding population are most valuable as parents of the next generation of offspring. The prediction model is additionally continuously rejuvenated as genotypical and phenotypic data from elite lines derived from the collaborating breeding programs is incorporated into the prediction models. In this manner, new germplasm may be infused into the system at any point. As lines derived from the recently infused germplasm advance within the breeding process, their genotypical and phenotypic data may be incorporated into the prediction models.

The purpose of phenotyping now is to pick the best lines from a segregating population and to judge fewer lines with larger replication in every cycle of selection. However during a GS driven breeding cycle, the aim of phenotyping is to estimate or re-estimate marker effects. It is far from clear at this point whether or not it will be advantageous to evaluate solely the best lines or to evaluate few lines with high replication. So separates the germplasm improvement cycle from the prediction model improvement cycle. Indeed, if we tend to use the rules for optimum QTL linkage mapping, analysis should include not just the best however the best and the the} worst lines Figure also emphasizes the requirement for model updating and re-evaluation. Marker effects might amendment as a results of allomorph frequency changes or of epistatic gene action. Model updating with every breeding cycle should mitigate reduced gains from GS caused by these mechanisms. Thus, GS may radically change the practice of field evaluation for breeders. Of course, despite the breeding technique used, final field evaluations of varieties across the target environments are going to be needed before they are distributed to farmers.

Accuracy declines as generation number between the last model update and selection candidates increases, because selection causes changes in variances, allele frequencies, and LD relationships between markers and QTL. Under random mating, simulations have shown model accuracy to decrease by about 5% per generation, but accuracy decrease was much more rapid under selection.

Factors Responsible for the Estimation Accuracy of GS Models

The response of genomic selection is that the output of varied factors responsible for estimation accuracy of GEBVs. These factors are interconnected in an exceedingly advanced and comprehensive manner. They include model performances, sample size and relatedness, marker density, gene effects, heritability and genetic design, and therefore the extent and distribution of linkage disequilibrium between markers and QTL.

Population Size

The most important characteristic of the population is its effective size. An obvious measure of population size is its census: how many individuals it contains. But populations with the same census size can behave quite differently. For a population of a given rate of inbreeding, the effective size is equal to the census size of a randomly mating ("ideal") population that would have that same rate.

Accuracy due to genetic relationships can represent from a small minority to a large majority proportion of the overall accuracy. The combination of long-distance LD due to pedigree relatedness (e.g., full sibs and half sibs) and short-distance ancestral LD due to small effective population size are among the key features of our training population. With improved marker technology, large TPs that use a representative sample of germplasm in a given breeding program may be a good strategy for long-term accuracy over a broad range of families. It has been observed to be monotonic increase in the prediction accuracy for grain yield with increasing population size without any substantial decrease in the slope. Studies in this the size of the training population is of crucial importance in genomic selection. The impact of the population size on the accuracy of genomic selection is less pronounced for fewer characters like grain moisture, which might be due to presence of larger variance among populations that can be efficiently utilized by few individuals per population. Parameters such as effective population size and QTL number strongly influence marker densities and TP sizes required for acceptable accuracy. Indeed, simulations similar to those of Meuwissen have shown that marker density needs to scale with effective population size. Until very low marker densities were reached, marker number had very little, if any, effect on prediction accuracies within families from various plant species. Likewise, GEBV accuracy of several traits in cattle, including net merit, was hardly affected when as many as 75% of the original markers were masked. Adequate marker density and TP size depend on QTL number and trait heritability. Calus and Veerkamp used the average r^2 between adjacent markers as a measure of marker density relative to decay of linkage disequilibrium. They found that for a highly heritability trait, average adjacent marker r^2 of 0.15 was sufficient, but for a low heritability trait, increasing the r^2 to 0.20 improved prediction accuracy. Heritability dramatically affects TP sizes required for successful GS, especially at h^2 less than 0.40.

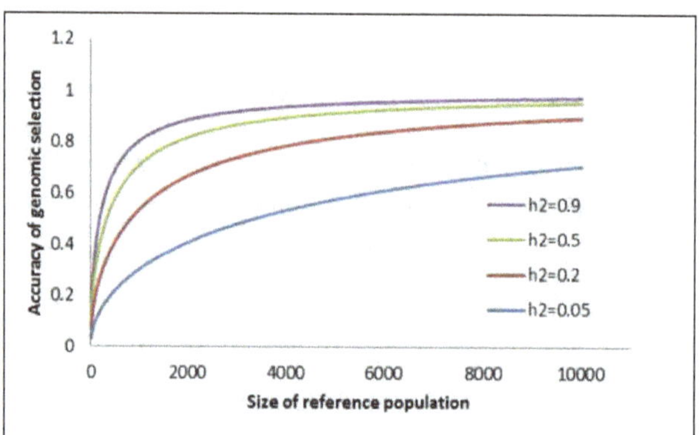

Relation between number of plants in the training population and accuracy of genomic selection for traits with different heritabilities.

Linkage Disequilibrium

Genetic drift is an important cause in generating LD, the non-independence of alleles at different loci. This non-independence allows marker alleles to predict the allelic state of nearby QTL, enabling marker genotypes to predict the phenotype. The LD intensity decays with greater distance between two markers. Decay rates which vary widely across species, populations, and genomes due to forces of mutation, recombination, population size, population mating marker density must increase with increases in Ne*c, where Ne is the effective population size and c is the recombination rate between loci. LD patterns. Marker density can be inferred by the rate of LD decay across the genome as inferred by the relationship b/w inter marker coefficient of determination r^2 and genetic distance. LD estimates can be used to determine target marker densities for GS at equilibrium, drift generating LD is balanced by recombination, causing it to decay, such that nearby loci are expected to be in higher LD than faraway loci. LD has a major effect on the operability of GS, so it has to be well understood before performing GS. LD is defined as the non-random association of alleles at different loci. It has been found that for high heritability trait average adjacent marker r^2 of 0.15 is sufficient but for low heritability trait increasing r^2 value to 0.2 improve accuracy of GEBV predictions.

Types of Marker Platforms

Since, then marker of choice is very important to accurate estimate GEBVBs, different platforms are available.

Genotyping by Sequencing

Advances in next generation technologies have driven the costs of DNA sequencing down to the point that genotyping-by sequencing (GBS) is now feasible for high diversity, large genome species. GBS is a highly multiplexed approach is based on high-throughput, next-generation sequencing of genomic subsets targeted by restriction enzymes (REs). Genotyping by sequencing (GBS) in any large genome species requires reduction of genome complexity. Genotyping-by-sequencing can be applied to different populations or even different species without any prior genomic knowledge as marker discovery is simultaneous with the genotyping of the population. GBS sequence allows access to any sequence within low copy genomic regions or regulatory regions controlling the expression of plant genes responsible for agronomically important phenotypes are often located in non-coding DNA. The use of GBS for GS, therefore, should be applicable to a range of model and non-model crop species to implement genomics-assisted breeding. Genotyping-by sequencing combines marker discovery and genotyping of large populations, making it a superb marker platform for breeding applications even within the absence of a reference genome sequence or previous polymorphism discovery. Additionally, the pliability and low price of GBS create this an ideal approach for genomics-assisted breeding.

Advantages of Genomic Selection

- The marker effects are calculable from the training population and used directly for GS within the involved breeding population, and QTL discovery, mapping, etc., are not needed.
- Each simulation and empirical studies reveal that GS produces larger gains per unit time than constitution selection. For instance, a simulation study in maize showed GS to be superior to MARS, notably for traits having low heritability. Further, GS will predict the performance of breeding lines additional accurately than that supported pedigree data, and GS appears to be an efficient tool for rising the potency of rice breeding.
- The selection index approach integrates appropriately weighted data from multiple traits into an index that is the premise for concurrent selection for the concerned traits. The genome-wide marker data is integrated into a range index either alone or in conjunction with phenotype data on one or additional traits. Simulation studies show that the on top of combined selection index approach of GS can increase the effectiveness of selection, considerably for low heritability traits.
- GS would tend to cut back the speed of inbreeding and also the loss of genetic variability as compared to selection based on breeding values calculable from phenotype data; this may be achieved while not sacrificing selection gains. This might be notably vital in species that show severe inbreeding depression.
- Genomic selection scheme consist of phenotyping for each selection cycle within the breeding population is not needed. This greatly reduces the length of breeding cycle, notably in perennial species. For instance, GS was calculable to reduce the selection cycle time from 19 years to simply 6 years just in case of oil palm (Elaeis guineensis). Further, GS was calculable to outperform MARS and phenotypic selection even with a population size of fifty once selection gain was considered on per unit time and price, however not on per selection cycle, basis. (The selection the choice) cycle is reduced as a result of GS does not need analysis of interbreeding performance of the plants being subjected to selection that is critical within the case of phenotypic selection. In perennial species, GS is anticipated to facilitate commercialization of improved genotypes at abundant shorter intervals of time than phenotypic selection.
- Genomic selection would possibly enable breeders to pick out parents for crossbreeding programs from among those lines that have not been evaluated within the target environment. This selection would be supported GEBVs of these lines estimated for their adaptation to the target environment. This could facilitate germplasm exchange and their utilization in breeding programs.

- Genotype X environment interaction could be a vital a part of phenotype and its estimation is sort of demanding. GS can utilize information on marker genotype and trait phenotype accumulated over time in varied analysis programs covering a variety of environments and integrate an identical in GEBV estimates of the various individuals/lines. This could enable GEBV estimation even for traits that they have never been tested.
- Theoretically, GEBV estimates is employed for the selection of parents for crossing programs and, possibly, for the development of hybrid varieties. These applications, however, ought to expect validation of the concept in apply.

Disadvantages of Genomic Selection

- GS has still not become popular plant breeding community primarily due to low evidence for its sensible utility. In fact, most discussions on its utility are for the most part statistical treatments and simulations that are not simply appreciated by plant breeders.
- The potential value of GS should be assessed with caution because GS has been mostly evaluated in simulation studies. There is an imperative have to be compelled to judge genomic selection in crop breeding situations to demonstrate its practical utility.
- Thc marker effects and, as a result, GEBV estimates would possibly modification attributable to changes in gene frequencies and epistatic interactions. This is often ready to necessitate amendment of the GS model with every breeding cycle therefore the gains from GS are not reduced.
- The accuracy of GEBV estimates has been evaluated exploitation simulation models based on additive genetic variance. These models ignore epistatic effects that does not seem to be realistic. It has been argued that since biological process makes alone a small contribution to the breeding value, the employment of solely additive genetic models for GS is additionally expected to maximize selection gains. However, this argument are planning to be entirely valid only for self-fertilizing species, where homozygous lines are used as parents as well as varieties.
- The selection response declines at a faster rate under GS than with pedigree selection. This may be reduced by continually together with new markers for the prediction of GEBVs. The long response under GS can also be raised by putting higher weights on the low-frequency favorable alleles, considerably within the start of GS program. GS is simpler than phenotypic selection on per unit time basis only if off-season/greenhouse facilities are accustomed grow up to three generations per annum. The utility and also the cost-effectiveness of GS would be uncertain wherever such facilities are not offered.

- The necessity for genotyping of an oversized variety of marker loci in every generation of selection adds considerably to the price of breeding programs. It has been projected that, inside the future, a bigger stress are going to be placed on the use of marker information than on composition information. Such a shift, however, would need the value of one marker information to be merely 1/5000 the price of phenotyping one entry.
- Implementation of GS would need intensive infrastructure and completely different resources, which might get on the so much aspect the reach of most moderate size public sector breeding programs, considerably within the developing countries. To boot, planning and execution of GS is kind of exigent, and additionally the breeders would be required to reorient their approach to the breeding programs.

Genome Editing

Novel genome editing tools, also referred to as genome editing with engineered nuclease (GEEN) technologies, allow cleavage and rejoining of DNA molecules in specified sites to successfully modify the hereditary material of cells. To this end, special enzymes such as restriction endonucleases and ligase can be used for cleaving and rejoining of DNA molecules in small genomes like bacterial and viral genomes. However, using restriction endonucleases and ligases, it is extremely difficult to manipulate large and complex genomes of higher organisms, including plant genomes. The problem is that the restriction endonucleases can only "target" relatively short DNA sequences. While such specificity is enough for short DNA viruses and bacteria, it is not sufficient to work with large plant genomes. The first efforts to create methods for the editing of complex genomes were associated with the designing of "artificial enzymes" as oligonucleotides (short nucleotide sequences) that could selectively bind to specific sequences in the structure of the target DNA and have chemical groups capable of cleaving DNA.

Targeted approach to address this challenge was the design of chimeric nucleases which are complex proteins containing one or two structural units, one of which catalyzes the cleavage of DNA, and the second is capable of selectively binding to specific nucleotide sequences of target molecule, providing the nuclease action to this site. These chimeric nucleases can be "produced" directly in the cell: to this end, appropriately engineered vectors encoding nucleases need to be introduced into cell. Such vectors are also supplied with nuclear localization signal which enables the nuclease to enter the cell nucleus thereby getting access to genomic DNA.

Zinc Finger Nucleases

ZFNs were the first generation of genome editing tools that use chimerically engineered nucleases which were developed after the discovery of the working principles of the functional Cys2-His2 zinc finger (ZF) domain. Each Cys2-His2 ZF domain consists of 30 amino acid residues, which are folded up to $\beta\beta\alpha$ configuration. Crystallographic structure analysis showed that the Cys2-His2 ZF proteins bind to DNA by inserting an α-helix of the protein into the major groove of the DNA-double helix. Each ZF protein has the ability to recognize 3 tandem nucleotides in the DNA. Generalized ZFN monomer consists of two different functional domains: artificial ZF Cys2-His2 domain at the N-terminal region and a nonspecific FokI DNA cleavage domain at the C-terminal region. FokI domain dimerization is critical for ZFN enzymatic activity. The observation that the modular recognition of zinc finger domains presents as a series to the corresponding, consecutive three bp targets enabled the realization that each of the individual zinc finger domains could be interchangeable and that the manipulation of the order of the domains would lead to unique binding specificities to the proteins harboring them thereby enabling targeting of specific, unique sequences in the genome. For example, a ZFN dimer, consisting of two 3 or 4 ZF domains, recognizes a target sequence of 18 or 24 base pairs, which statistically form unique sites in the genomes of most organisms.

The design and application of ZFNs involve modular design, assembly, and optimization of zinc fingers against specific target DNA sequences followed by linking of individual ZFs towards targeting larger sequences. Over the years, zinc finger domains have been generated to recognize a large number of triplet nucleotides. This enabled the selection and linking of zinc fingers in a sequence that would permit recognition of the target sequence of interest.

Since the first report on zinc fingers in 1996, they have been successfully used in several organisms including plants. Examples include targeted inactivation of endogenous genes in Arabidopsis, high frequency modification of tobacco genes, and precise targeted addition of a herbicide-tolerance gene as well as insertional disruption of a target locus in maize. ZFNs have also been used for trait stacking in maize.

Zinc finger nucleases have revolutionized the field of genome editing by demonstrating the ability to manipulate genomic sites of interest and opened the gates for both basic and applied research. ZFNs provide advantages over other tools with respect to efficiency, high specificity, and minimal Non-target effects and current efforts are focused on further improving design and delivery as well as expanding their applications in diverse crops of interest.

Transcription Activator-like Effector Nucleases

The quest for efficient and selective manipulation of target genomic DNA led to the identification of unique transcription activator-like effector (TALE) proteins that recognize and activate specific plant promoters through a set of tandem repeats that formed the basis for the creation of a new genome editing system consisting of chimeric nucleases called TALE nucleases (TALENs). TALE proteins consist of a central domain responsible for DNA binding, nuclear localization signal, and a domain that serves as activator of transcription of the target gene.

It is shown that the DNA-binding domain in TALE monomers in turn consists of a central repeat domain (CRD) that confers DNA binding and host specificity. The CRD consists of tandem repeats of 34 amino acid residues and each 34-amino acid long repeat in the CRD binds to one nucleotide in the target nucleotide sequence. Two of the amino acids of the repeat, located at positions 12 and 13, are highly variable (repeat variable diresidue (RVD)) and are responsible for the recognition of specific nucleotide with degeneracy of binding several nucleotides with differential efficiency. The last tandem repeat binding to nucleotide at the 3′-end of the recognition site consists of 20 amino acid residues only and, therefore, it is named as half-repeat. While TALE proteins, in general, can be designed to bind any DNA sequence of interest, studies have demonstrated that the 5′-most nucleotide base of the DNA sequence bound by a TALE protein should always be a Thymidine and that a deviation from this requirement can affect the efficacy of TALE transcription factors (TALE-TF), TALE recombinases (TALE-R), and TALENs.

After the DNA code recognition requirements by TALE proteins have been cracked, the very first effort undertaken was the creation of chimeric TALEN nucleases. For this purpose, the sequence encoding the DNA-binding TALE domain was inserted into a plasmid vector previously used to create ZFN. This resulted in the creation of a synthetic, chimeric sequence-specific nuclease genetic construct containing the DNA-binding domain of TALEs and the catalytic domain of FokI restriction endonuclease. This construct helped to create artificial nucleases with DNA-binding domain and different RVDs that can target any nucleotide sequence of interest.

In most studies, the monomers with RVDs Asn and Ile (NI), Asn and Gly (NG), two Asn (NN), and His and Asp (HD) bind to nucleotides A, T, G, and C, respectively. NN, the most common RVD that specifies G, was also found to bind to A. This suboptimal or lack of specificity is a concern for the use of engineered TALEs for targeting DNA. Another RVD NK has less functional efficiency compared to NN, although it has demonstrated guanine specificity. Several studies have also shown that the use of NH or NK RVDs for specific binding of guanine reduces the risk of Non-target effects. It has been shown that in RVD (NI, NG, NN, or HD) the first amino acid residue, whether it is N or H, is responsible for the stabilization of spatial conformation although it does not directly bind to a nucleotide, whereas the second amino acid residue binds to a nucleotide either through hydrogen bonding with nitrogenous bases (in case of D and N amino acids) or through van der Waals forces (in case of I and G).

Based on the mode of action and specificity of TALENs, it should be possible to introduce double strand breaks in any location of the genome as long as that location harbors the recognition sequence corresponding to the DNA-binding domains of TALENs.

There is another condition that also needs to be met, that is, the requirement of the presence of Thymidine before the 5′ end of the intended target sequence since it has been demonstrated that the W232 residue in the N-terminal portion of the DNA-binding domain interacts with the Thymidine and influences the binding efficiency. It is also possible to overcome this 5′ Thymidine constraint by developing mutant variants of TALEN N-terminal domain which can bind other nucleotides. Considering the ease of site-directed manipulation using TALEN system, within a short period of time after the unraveling of the TALEN mode of action, the genes modified by this system have been used successfully in several animal and plant species and the plant examples include rice, wheat, Arabidopsis, potato, and tomato.

Oligonucleotide-directed Mutagenesis

After first successful exploitation in mammalian systems, oligonucleotide-directed mutagenesis (ODM) has become another novel gene editing tool for plants. ODM, a tool for targeted mutagenesis, uses a specific 20- to 100-base long oligonucleotide, the sequence of which is identical to the target sequence in the genome except that it contains a single base pair change (intended mutation to be inserted in the genome) towards achieving site-directed editing of gene/sequence of interest. When these synthetic oligonucleotides or repair templates with homology to a specific region of the target gene are transiently exposed to the plant cells by using a variety of specific delivery methods, they bind to the targets and activate cell's natural repair machinery which recognizes the single mismatch in the template and then copies that mismatch or mutation into the target sequence through repair process. This produces the desired targeted single nucleotide or base editing in the plant genome that confers novel function or trait while the plant cell degrades the repair template oligonucleotide. Using tissue culture methods, cells with edited sequences are subsequently regenerated and genome edited novel varieties with improved traits/characteristics are developed through traditional breeding.

Clustered Regularly Interspaced Short Palindromic Repeats

Another novel genome editing system that has emerged recently and has become widely popular is the clustered regularly interspaced short palindromic repeats (CRISPR)/CRISPR associated (Cas) protein system with the most prominent being the CRISPR/Cas9 (based on Cas9 protein). This is a method that utilizes adaptive bacterial and archaeal immune system, the mechanism of which relies on the presence of special sites in the bacterial genome called CRISPR loci. These loci are composed of operons encoding the Cas9 protein and a repeated array of repeat spacer sequences. The spacers in the repeat array are short fragments that are derived from foreign DNA (viral or plasmid) that have become integrated into bacterial genome following recombination.

Unlike the chimeric TALEN proteins, target site recognition by CRISPR/Cas9 system is accomplished by the complementary sequence based interaction between the guide (noncoding) RNA and DNA of the target site and the guide RNA and Cas protein complex has the nuclease activity for exact cleavage of double-stranded DNA using Cas9 endonuclease.

The most "popular" system is the CRISPR/Cas type II-A system found in the bacterium Streptococcus pyogenes and composed of three genes encoding CRISPR RNA (crRNA), trans-activating crRNA (tracrRNA), and Cas9 protein. Based on this system, universal genetic constructs encoding artificial elements of CRISPR/Cas "genome editor" have been created. Also, a simplified version of the system, functioning as a complex of Cas9 protein and a single guide RNA, consisting of CRISPR tracrRNA and short, mature crRNA was created. The guide sequence identifies the target DNA site and binds to it based on complementarity and Cas9 cleaves the DNA in target point.

CRISPR system can be used for the creation of genetically modified cells grown in culture and living organisms. In the first case, plasmids or viral vectors which provide high and stable synthesis of CRISPR/Cas9 system elements are introduced into cells. In the second case, cultured protoplasts and a plasmid coding CRISPR/Cas elements are used to obtain genetically modified plants. Another approach, applied for plants, is the use of Agrobacterium, the natural "genetic engineer", that contains a special plasmid harboring CRISPR/Cas9 system.

Thus, due to its simplicity, efficiency, and wide capabilities, in a short time CRISPR/Cas9 system has already found use in various fields of fundamental and applied biology, biotechnology, and genetic engineering.

Repair of Cleaved Genomic Sites

An important step in the genome editing process is the repair of the DNA break created by the nucleases. DNA break gets repaired by the endogenous cellular mechanisms: nonhomologous end-joining (NHEJ) or homology-dependent (or directed) repair (HDR). NHEJ is the simplest mechanism where the ends of the cleaved DNA are joined together, often resulting in the insertion or deletion of nucleotides (indels) thereby shifting the gene reading frame, resulting in a gene "knockout". If indels are not observed, the DNA is recovered, and there are no noticeable changes. On the other hand, HDR is a mechanism where a sequence containing homology to target is used as a template for repairing the break or the DNA lesion. Therefore, by providing a template that contains a desired sequence of interest flanked by sequences homologous to both sides of the break point, one can force the insertion of that desired sequence into the target site. When HDR occurs, a homologous recombination is used to enable new sequences for gene recovery or insertion. This method is simple, provides the exact impact on DNA target, and can be used at almost any modern molecular biology laboratory.

Practical Applications of Genome Editing Systems

Application of Genome Editors for Functional Genomics

Several different types of genome modifications can be achieved by utilizing ZFN, TALEN, ODM, and CRISPR/Cas genome editing systems. These include creation of point mutations, insertion of new genes in specific locations or deletion of large regions of the nucleotide sequences, and correction or substitution of individual genetic elements and gene fragments.

While introducing modifications to various genomic elements in plant cells and examining the results, scientists were able to investigate the role of individual genes in the functioning of individual cells and the organism as a whole. For example, the unique ability of CRISPR/Cas9 system to selectively bind to specific DNA sites has helped to regulate gene activity. For this purpose, proteins activating or repressing the activity of promoters that control the gene function can be attached to the catalytically inactive mutant Cas9 protein. In one example, it was shown that complex binding to the target DNA can inhibit or stimulate the function of the target gene.

Furthermore, using CRISPR/Cas9 system, several genetic constructs targeted to different genome sites can simultaneously be introduced into cells. This is a welcome feature in investigating intergenic interaction, if any, because several genes are simultaneously affected by the CRISPR/Cas9 system. For example, using this approach, it was possible to identify genes involved in crop domestication process.

Application of Genome Editing Systems in Crop Improvement

Genome editing technologies have wide practical applications for solving one of the most important tasks of modern biotechnology—the creation of new varieties of crops, which are high-yielding and resistant to abiotic and biotic stresses and also have high nutritional value. To this end, genome editing system has been used in plant breeding (1) to insert point mutations similar to natural SNPs, (2) to make small modifications to gene function, (3) for integration of foreign genes, (4) for gene pyramiding and knockout, and (5) for the repression or activation of gene expression, as well as (6) epigenetic editing.

For example, the use of ZFN in Arabidopsis thaliana and Zea mays has led to the successful development of herbicide tolerant genotypes through insertion of herbicide-resistance genes into targeted sites in the genome. ZFN was also used for the targeted modification of an endogenous malate dehydrogenase (MDH) gene in plants and the plants containing modified MDH have shown increased yield. ODM technique has been significantly advanced through Cibus Rapid Trait Development System (RTDS) and this technology has been successfully applied in several crops. Applications include but are not limited to the generation of herbicide tolerance, insect resistance, enhanced disease resistance (bacterial and viral), improved nutritional value, and enhanced yield without the introduction of foreign genes as has been used in traditional genetic engineering approach for crop development. A precise editing of CAC to TAC using ODM RTDS technology has been demonstrated that converts BEP to GFP by changing Histidine (H66) to Tyrosine (Y66) in GFP protein. This approach has offered a nontransgenic breeding tool for crops.

Using the CRISPR/Cas9 technology, Jiang have obtained "a biotech" oil fromCamelina sativa seeds with an improved fatty acid composition, which makes it more beneficial to human health, more resistant to oxidation, and more appropriate for the production of certain commercial chemicals including biofuels. Soyk used targeted mutagenesis of SP5G gene of tomato to create plants with rapid flowering and more compact bush, which in turn resulted in earlier harvest. In another effort, Osakabe , using the CRISPR-induced mutagenesis of OST2 gene in Arabidopsis, were able to obtain new alleles that confer salt stress resistance to plants.

Modulation of the gibberellin biosynthesis by genome editing methods has allowed creation of dwarf fruit trees which have great potential for increasing productivity through higher density plantings and reduced labor costs. This results in a reduction of land, water, pesticide, and fertilizer use. In addition, genome editing for inhibition of ethylene biosynthesis, which plays a very important role in fruit ripening process or its signaling pathways, enables creation of new varieties with extended shelf life.

A major area of application of genome editing approaches in plant breeding is to create varieties resistant to various pathogens and pests. These methods have been used for the modification of the key plant immunity stages at different levels in several crops. This goal can be achieved by modifying (1) susceptibility genes (S-genes), (2) resistance genes (R-genes), (3) genes regulating the interaction between the effector and target, and (4) the genes regulating plant hormonal balance. For example, wheat genotypes resistant to powdery mildew disease were obtained by TALEN- and CRISPR/Cas9-mediated genome editing on mildew-resistance locus O (MLO). Genome editing technologies have also been used to produce plants resistant to bacterial leaf blight, caused by Xanthomonas oryzae pv. oryzae.

The CRISPR/Cas9 system has been investigated for its efficacy in providing interference against geminiviruses by using a transient transformation system such that N. benthamiana degradation/suppression of curly top virus genome by single guide RNA/Cas9 (sgRNA/Cas9) has been demonstrated. In other efforts, where sgRNAs specific for tomato yellow leaf curl virus (TYLCV) or bean yellow dwarf virus (BeYDV) sequences were introduced into N. benthamiana plants expressing Cas9 endonuclease and challenged with the corresponding viruses, it was demonstrated that the CRISPR/Cas9 system not only targeted viruses for degradation but also introduced mutations at the target sequences due to interference with the copy number of freely replicating viruses.

Metabolic pathways that regulate hormonal balance can also be modified using the genome editing technologies to enhance the immunomodulatory component of the plants immune system. This can be achieved by deactivating the ethylene-responsive factor (ERF). In particular, ethylene-dependent pathway in rice has been successfully modified by CRISPR/Cas9-mediated target OsERF922 gene mutations, resulting in increased resistance to Magnaporthe oryzae.

CRISPR/Cas9 has been used to knock out eIF4E gene that encodes the eukaryotic translation initiation factor essential for translation of viruses, in Cucumis sativus, and that knockout confers resistance to viruses such as cucumber vein yellowing virus (CVYV), zucchini yellow mosaic virus (ZYMV), and papaya ring spot mosaic virus-W (PRSV-W). In addition, CRISPR/Cas9 was demonstrated to be an efficient system for rapid and efficient genome editing in Phytophthora sojae, an oomycete pathogen of Soybean, by modifying the pathogenicity gene (Avr4/6), thereby opening up an avenue for the much needed functional genomics work in Phytophthora sojae towards the ultimate goal of controlling this pathogen.

Similarly, existing genome editing methods, in particular, CRISPR/Cas9 method, have been successfully used to obtain plants resistant to herbicides. For example, editing of ALS2 gene in maize (acetolactate synthase or ALS is a key enzyme in the biosynthesis of amino acids in plants and has been inhibited by sulfonylurea herbicides) allowed the creation of a mutant corn plant resistant to chlorsulfuron.

Another interesting area of biotechnology where CRISPR/Cas9 system has significant application is the development of plants capable of synthesizing human proteins such as insulin, necessary for patients with diabetes mellitus, or albumin, which is used in the treatment of hemorrhagic shock, burns, hypoproteinemia, and cirrhosis. At present, albumin is prepared from human plasma which is in a very limited supply; however, global demand for albumin is constantly growing and currently is equal to 500 tons per year. To meet the growing needs human albumin gene is already introduced into rice genome using genomic engineering techniques. Such expressed proteins can be isolated from plant and animal tissues, where it is synthesized, and after clarification, it can be used for medical purposes.

These novel genome editing techniques are being widely used for the purpose of crop improvement including new bioenergy crop developments. However, the use of tissue culture with these GEEN methods may also create complexities that could slow the process of genome editing.

Safety Assessment Aspects of Genome Editing Systems

Non-target Effects

Genome editing techniques, in essence, preserve the native genomic structure and, therefore, are considered as a safe technology for crop improvement. Despite this general understanding, there are some concerns related to the biosafety of crops created using these methods. One main concern in terms of its biosafety is the possibility of Non-target effects of synthetic nucleases during genome editing.

During the biotechnological application of genome editing methods, efficiency and specificity of the engineered nucleases are the two most important functional requirements and are closely related to the choice of the target site. For each endogenous genomic locus, efficiency of DNA cleavage (both target and Non-target) depends not only on the nuclease activity (such as FokI domains and Ruv domains of the Cas9 proteins), but also on the availability of a target site and affinity of the DNA-binding domain (e.g., TAL effector domains and guide RNA, gRNA) to the target sequence. Specificity of engineered nucleases largely depends on the binding affinity of nuclease-DNA, including the binding of zinc finger to DNA (ZFNs), TAL effector to DNA (TALENs), and hybridization of gRNA with DNA (CRISPR), although dimerization of FokI domain (ZFNs and TALENs) and Cas9 interaction with the motif contiguous to protospacer adjacent motif (PAM) may also play an important role. In case of ZFNs, while examples abound with respect to the binding efficiency of canonical C_2H_2 binding domain containing ZFNs, investigations on the utility of noncanonical ZFNs such as those containing C_3H_1 binding domain have demonstrated high levels of binding efficiency.

To minimize Non-target effects of genome editing systems, a crucial aspect is the careful selection of sites for the introduction of the double-stranded breaks by performing a prior bioinformatics analysis. When choosing the desired sites, sites of repeated sequences and sites having a high homology with other regions of the genome should be avoided. In this regard, to facilitate the selection of the target sites for nucleases and experimental verification of the presence of Non-target effects, several software packages were developed that enable nuclease design and validation.

Regulation of Plants Created by Genome Editing

The novel genome editing systems help to introduce stably inherited point modifications into the plant genome, and transgenic region can be easily removed after editing a target gene. This allows creation of nontransgenic plants and improved crop varieties. These technologies are faster compared to traditional breeding methods and help to obtain the null segregant lines that have lost the transgene insertion. Plants with targeted mutations developed by genome editing technology are nearly identical to plants obtained by classical breeding, and their safety must be assessed taking into account the resulting product rather than the process used to create them. In this context, ODM-derived products are in many cases indistinguishable from conventionally bred or traditional mutagenesis products; therefore, such products should not be regulated in the same way as the products generated by genetic engineering methods. Using CRISPR-Cas9 system, it becomes possible to obtain marker-free genetically engineered crops, that is, without marker genes of antibiotic resistance. Thus, in the case of new varieties with targeted mutations, developed using genome editing systems, the existing operating rules for the regulation of genetically modified plants should not be applied. Currently genome editing technologies are being discussed by various advisory and regulatory authorities in the context of GMO legislation. Cultures and plants obtained using genome editing techniques are considered as nongenetically modified. The European Commission is expected to publish a report on regulatory uncertainty of genome editing methods.

Multitude of Advantages and Perspectives of GEENs

Tools of genome editing have a significant impact on basic and applied research in plant biology. The simplified approach to gene/genome editing represents a valuable tool for plant researchers in functional analysis of genes and for breeders in the integration of key genes in the genomes of agriculturally important crops. Genome editing systems have several attractive features including simplicity, efficiency, high specificity, minimal Non-target effects, and amenability to multiplexing and thus are very promising for use in plant breeding.

Site-directed mutagenesis of different genes can provide important information about their functions. Simultaneous targeting of multiple genes/loci by applying multiplex strategies can promote research to identify the role of individual genes in the intracellular signaling pathways and aid in the engineering of complex, multigenic agronomic traits in crops. The preferred use of CRISPR-Cas9 system can be exemplified in completely knockout gene function, microRNA knockdown screening, and programmed editing of certain loci by genome editing systems that can provide a functional separation of cis- and trans-regulatory elements/factors with high accuracy. Another prospective application of CRISPR-Cas9 system may be its use in the formation of conditional alleles, providing spatial and temporal control of gene expression to study the function of lethal genes. Use of inducible or tissue-specific promoters for expression of Cas9 and sgRNA can be instrumental for gene expression regulation in a specific tissue, in development stage, or in different environmental conditions.

CRISPR-Cas system opens up wide possibilities for labeling endogenous genes with fluorescent proteins to visualize their expression in vivo. Using fluorescent labeled dCas9, changes of genome dynamics/chromosome architectural changes during plant development and their response to environmental stimuli can be learned. These technologies can also be used for the selection of the specific cell types that greatly facilitate the study of various functional aspects. Use of dCas9 can provide a new platform for the selection of activation/repression effector domains to specific genomic loci for regulating endogenous gene expression.

In addition, these technologies can be successfully used in the work on epigenome editing via the selection of proteins responsible for histone modification and DNA methylation, which has emerged as a new way of regulating cellular functions in plants. For the purpose of understanding epigenetic regulation, CRISPR-Cas9 system can also be used for the enrichment of chromatin target sites for the identification of proteins attached to enriched chromatin. Likewise, CRISPR-Cas9 can be used as a tool to identify regulatory proteins binding to specific DNA sequences controlling the expression of genes.

Plant Tissue Culture

Tissue culture is the in vitro aseptic culture of cells, tissues, organs or whole plant under controlled nutritional and environmental conditions often to produce the clones of plants. The resultant clones are true-to type of the selected genotype. The controlled conditions provide the culture an environment conducive for their growth and multiplication. These conditions include proper supply of nutrients, pH medium, adequate temperature and proper gaseous and liquid environment.

Plant tissue culture technology is being widely used for large scale plant multiplication. Apart from their use as a tool of research, plant tissue culture techniques have in recent years, become of major industrial importance in the area of plant propagation, disease elimination, plant improvement and production of secondary metabolites. Small pieces of tissue (named explants) can be used to produce hundreds and thousands of plants in a continuous process. A single explant can be multiplied into several thousand plants in relatively short time period and space under controlled conditions, irrespective of the season and weather on a year round basis. Endangered, threatened and rare species have successfully been grown and conserved by micropropagation because of high coefficient of multiplication and small demands on number of initial plants and space.

In addition, plant tissue culture is considered to be the most efficient technology for crop improvement by the production of somaclonal and gametoclonal variants. The micropropagation technology has a vast potential to produce plants of superior quality, isolation of useful variants in well-adapted high yielding genotypes with better disease resistance and stress tolerance capacities. Certain type of callus cultures give rise to clones that have inheritable characteristics different from those of parent plants due to the possibility of occurrence of somaclonal variability, which leads to the development of commercially important improved varieties. Commercial production of plants through micropropagation techniques has several advantages over the traditional methods of propagation through seed, cutting, grafting and air-layering etc. It is rapid propagation processes that can lead to the production of plants virus free. Coryodalisyanhusuo, an important medicinal plant was propagated by somatic embryogenesis from tuber-derived callus to produce disease free tubers. Meristem tip culture of banana plants devoid from banana bunchy top virus (BBTV) and brome mosaic virus (BMV) were produced. Higher yields have been obtained by culturing pathogen free germplasmin vitro. Increase in yield up to 150% of virus-free potatoes was obtained in controlled conditions.

In plant cell culture, plant tissues and organs are grown in vitro on artificial media, under aseptic and controlled environment. The technique depends mainly on the concept of totipotentiality of plant cells which refers to the ability of a single cell to express the full genome by cell division. Along with the totipotent potential of plant cell, the capacity of cells to alter their metabolism, growth and development is also equally important and crucial to regenerate the entire plant. Plant tissue culture medium contains all the nutrients required for the normal growth and development of plants. It is mainly composed of macronutrients, micronutrients, vitamins, other organic components, plant growth regulators, carbon source and some gelling agents in case of solid medium. Murashige and Skoog medium (MS medium) is most extensively used for the vegetative propagation of many plant species in vitro. The pH of the media is also important that affects both the growth of plants and activity of plant growth regulators. It is adjusted to the value between 5.4 - 5.8. Both the solid and liquid medium can be used for culturing. The composition of the medium, particularly the plant hormones and the nitrogen source has profound effects on the response of the initial explant.

Plant growth regulators (PGR's) play an essential role in determining the development pathway of plant cells and tissues in culture medium. The auxins, cytokinins and gibberellins are most commonly used plant growth regulators. The type and the concentration of hormones used depend mainly on the species of the plant, the tissue or organ cultured and the objective of the experiment. Auxins and cytokinins are most widely used plant growth regulators in plant tissue culture and their amount determined the type of culture established or regenerated. The high concentration of auxins generally favors root formation, whereas the high concentration of cytokinins promotes shoot regeneration. A balance of both auxin and cytokinin leads to the development of mass of undifferentiated cells known as callus.

Maximum root induction and proliferation was found in Stevia rebaudiana, when the medium is supplemented with 0.5 mg/l NAA. Cytokinins generally promote cell division and induce shoot formation and axillary shoot proliferation. High cytokinin to auxin ratio promotes shoot proliferation while high auxin to cytokinins ratio results in root formation. Shoot initiation and proliferation was found maximum, when the callus of black pepper was shifted to medium supplemented with BA at the concentration of 0.5 mg/l. Gibberellins are used for enhanced growth and to promote cell elongation. Maximum shoot length was observed in Phalaenopsis orchids when cultured in medium containing 0.5 mg/l GA_3.

Tissue Culture in Agriculture

As an emerging technology, the plant tissue culture has a great impact on both agriculture and industry, through providing plants needed to meet the ever increasing world demand. It has made significant contributions to the advancement of agricultural sciences in recent times and today they constitute an indispensable tool in modern agriculture.

Biotechnology has been introduced into agricultural practice at a rate without precedent. Tissue culture allows the production and propagation of genetically homogeneous, disease-free plant material. Cell and tissue in vitro culture is a useful tool for the induction of somaclonal variation. Genetic variability induced by tissue culture could be used as a source of variability to obtain new stable genotypes. Interventions of biotechnological approaches for in vitro regeneration, mass micropropagation techniques and gene transfer studies in tree species have been encouraging. In vitro cultures of mature and immature zygotic embryos are applied to recover plants obtained from inter-generic crosses that do not produce fertile seeds. Genetic engineering can make possible a number of improved crop varieties with high yield potential and resistance against pests. Genetic transformation technology relies on the technical aspects of plant tissue culture and molecular biology for:

- Production of improved crop varieties.
- Production of disease-free plants (virus).
- Genetic transformation.
- Production of secondary metabolites.
- Production of varieties tolerant to salinity, drought and heat stresses.

Germplasm Conservation

In vitro cell and organ culture offers an alternative source for the conservation of endangered genotypes. Germplasm conservation worldwide is increasingly becoming an essential activity due to the high rate of disappearance of plant species and the increased need for safeguarding the floristic patrimony of the countries. Tissue culture protocols can be used for preservation of vegetative tissues when the targets for conservation are clones instead of seeds, to keep the genetic background of a crop and to avoid the loss of the conserved patrimony due to natural disasters, whether biotic or abiotic stress. The plant species which do not produce seeds (sterile plants) or which have 'recalcitrant' seeds that cannot be stored for long period of time can successfully be preserved via in vitro techniques for the maintenance of gene banks.

Cryopreservation plays a vital role in the long-term in vitro conservation of essential biological material and genetic resources. It involves the storage of in vitro cells or tissues in liquid nitrogen that results incryo-injury on the exposure of tissues to physical and chemical stresses. Successful cryopreservation is often ascertained by cell and tissue survival and the ability to re-grow or regenerate into complete plants or form new colonies. It is desirable to assess the genetic integrity of recovered germplasm to determine whether it is 'true-to-type' following cryopreservation. The fidelity of recovered plants can be assessed at phenotypic, histological, cytological, biochemical and molecular levels, although, there are advantages and limitations of the various approaches used to assess genetic stability. Cryobionomics is a new approach to study genetic stability in the cryopreserved plant materials. The embryonic tissues can be cryopreserved for future use or for germplasm conservation.

Embryo Culture

Embryo culture is a type of plant tissue culture that is used to grow embryos from seeds and ovules in a nutrient medium. In embryo culture, the plant develops directly from the embryo or indirectly through the formation of callus and then subsequent formation of shoots and roots. The technique has been developed to break seed dormancy, test the vitality of seeds, production of rare species and haploid plants. It is an effective technique that is employed to shorten the breeding cycle of plants by growing excised embryos and results in the reduction of long dormancy period of seeds. Intra-varietal hybrids of an economically important energy plant "Jatropha" have been produced successfully with the specific objective of mass multiplication. Somatic embryogenesis and plant regeneration has been carried out in embryo cultures of Jucara Palm for rapid cloning and improvement of selected individuals. In addition, conservation of endangered species can also be attained by practicing embryo culture technique. Recently a successful protocol has been developed for the in vitro propagation of Khayagrandifoliolaby excising embryos from mature seeds. The plant has a high economic value for timber wood and for medicinal purposes as well. This technique has an important application in forestry by offering a mean of propagation of elite individuals where the selection and improvement of natural population is difficult.

Genetic Transformation

Genetic transformation is the most recent aspect of plant cell and tissue culture that provides the mean of transfer of genes with desirable trait into host plants and recovery of transgenic plants. The technique has a great potential of genetic improvement of various crop plants by integrating in plant biotechnology and breeding programmes. It has a promising role for the introduction of agronomically important traits such as increased yield, better quality and enhanced resistance to pests and diseases.

Genetic transformation in plants can be achieved by either vector-mediated (indirect gene transfer) or vectorless (direct gene transfer) method. Among vector dependant gene transfer methods, Agrobacterium-mediated genetic transformation is most widely used for the expression of foreign genes in plant cells. Successful introduction of agronomic traits in plants was achieved by using root explants for the genetic transformation. Virus-based vectors offers an alternative way of stable and rapid transient protein expression in plant cells thus providing an efficient mean of recombinant protein production on large scale.

Recently successful transgenic plants of Jatropha were obtained by direct DNA delivery to mature seed-derived shoot apices via particle bombardment method. This technology has an important impact on the reduction of toxic substances in seeds thus overcoming the obstacle of seed utilization in various industrial sector. Regeneration of disease or viral resistant plants is now achieved by employing genetic transformation technique. Researchers succeeded in developing transgenic plants of potato resistant to potato virus Y (PVY) which is a major threat to potato crop worldwide. In addition, marker free transgenic plants of Petunia hybrida were produced using multi-auto-transformation (MAT) vector system. The plants exhibited high level of resistance to Botrytis cinerea,causal agent of gray mold.

Protoplast Fusion

Somatic hybridization is an important tool of plant breeding and crop improvement by the production of interspecific and intergeneric hybrids. The technique involves the fusion of protoplasts of two different genomes followed by the selection of desired somatic hybrid cells and regeneration of hybrid plants. Protoplast fusion provides an efficient mean of gene transfer with desired trait from one species to another and has an increasing impact on crop improvement. Somatic hybrids were produced by fusion of protoplasts from rice and ditch reed using electrofusion treatment for salt tolerance.

In vitro fusion of protoplast opens a way of developing unique hybrid plants by overcoming the barriers of sexual incompatibility. The technique has been applicable in horticultural industry to create new hybrids with increased fruit yield and better resistance to diseases. Successful viable hybrid plants were obtained when protoplasts from citrus were fused with other related citrinae species. The potential of somatic hybridization in important crop plants is best illustrated by the production of intergeneric hybrid plants among the members of Brassicaceae. To resolvethe problem of loss of chromosomes and decreased regeneration capacity, successful protocol has been established for the production of somatic hybrid plants by using two types of wheat protoplast as recipient and protoplast of Haynaldiavillosa as a fusion donor. It is also employed as an important gene source for wheat improvement.

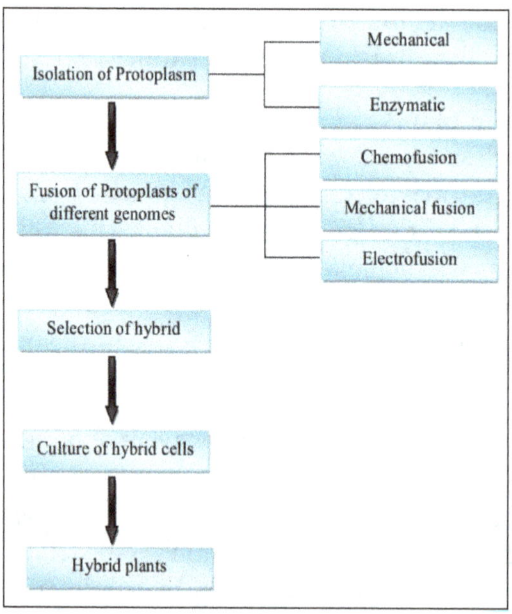

Production of hybrid plant via protoplast fusion.

Haploid Production

The tissue culture techniques enable to produce homozygous plants in relatively short time period through the protoplast, anther and microspore cultures instead of conventional breeding.

Haploids are sterile plants having single set of chromosomes which are converted into homozygous diploids by spontaneous or induced chromosome doubling. The doubling of chromosomes restores the fertility of plants resulting in production of double haploids with potential to become pure breeding new cultivars. The term androgenesis refers to the production of haploid plants from young pollen cells without undergoing fertilization. Sudherson reported haploid plant production of sturt's desert pea by using pollen grains as primary explants. The haploidy technology has now become an integral part of plant breeding programs by speeding up the production of inbred lines and overcoming the constraints of seed dormancy and embryo non-viability. The technique has a remarkable use in genetic transformation by the production of haploid plants with induced resistance to various biotic and abiotic stresses. Introduction of genes with desired trait at haploid state followed by chromosome doubling led to the production of double haploids inbred wheat and drought tolerant plantswere attained successfully.

Techniques of Plant Tissue Culture

Micropropagation

Micropropagation starts with the selection of plant tissues (explant) from a healthy, vigorous mother plant. Any part of the plant (leaf, apical meristem, bud and root) can be used as explant.

Stage 0: Preparation of Donor Plant

Any plant tissue can be introduced in vitro. To enhance the probability of success, the mother plant should be ex vitro cultivated under optimal conditions to minimize contamination in the in vitro culture.

Stage I: Initiation Stage

In this stage an explant is surface sterilized and transferred into nutrient medium. Generally, the combined application of bactericide and fungicide products is suggested. The selection of products depends on the type of explant to be introduced. The surface sterilization of explant in chemical solutions is an important step to remove contaminants with minimal damage to plant cells. The most commonly used disinfectants are sodium hypochlorite, calcium hypochlorite, ethanol and mercuric chloride ($HgCl_2$). The cultures are incubated in growth chamber either under light or dark conditions according to the method of propagation.

Stage II: Multiplication Stage

The aim of this phase is to increase the number of propagules. The number of propagules is multiplied by repeated subcultures until the desired (or planned) number of plants is attained.

Stage III: Rooting Stage

The rooting stage may occur simultaneously in the same culture media used for multiplication of the explants. However, in some cases it is necessary to change media, including nutritional modification and growth regulator composition to induce rooting and the development of strong root growth.

Stage IV: Acclimatization Stage

At this stage, the in vitro plants are weaned and hardened. Hardening is done gradually from high to low humidity and from low light intensity to high light intensity. The plants are then transferred to an appropriate substrate (sand, peat, compost etc.) and gradually hardened under greenhouse.

Cisgenesis

Cisgenesis is a product designation for a category of genetically engineered plants. A variety of classification schemes have been proposed that order genetically modified organisms based on the nature of introduced genotypical changes, rather than the process of genetic engineering.

Potatoes after treatment with Phytophthora infestans. The normal potatoes have blight but the cisgenic potatoes are healthy.

Cisgenesis (from "same" and "beginning") is one term for organisms that have been engineered using a process in which genes are artificially transferred between organisms that could otherwise be conventionally bred. Genes are only transferred between closely related organisms. Nucleic acid sequences must be isolated and introduced using the same technologies that are used to produce transgenic organisms, making cisgenesis similar in nature to transgenesis. The term was first introduced in 2000 by Henk J. Schouten and Henk Jochemsen, and in 2004 a PhD thesis by Jan Schaart of Wageningen University in 2004, discussing making strawberries less susceptible to Botrytis cinerea.

In Europe, currently, this process is governed by the same laws as transgenesis. While researchers at Wageningen University in the Netherlands feel that this should be changed and regulated in the same way as conventionally bred plants, other scientists, writing in Nature Biotechnology, have disagreed. In 2012 the European Food Safety Authority (EFSA) issued a report with their risk assessment of cisgenic and intragenic plants. They compared the hazards associated with plants produced by cisgenesis and intragenesis with those obtained either by conventional plant breeding techniques or transgenesis. The EFSA concluded that "similar hazards can be associated with cisgenic and conventionally bred plants, while novel hazards can be associated with intragenic and transgenic plants".

Cisgenesis has been applied to transfer of natural resistance genes to the devastating disease Phytophthora infestans in potato and scab (Venturia inaequalis) in apple.

Cisgenesis and transgenesis use artificial gene transfer, which results in less extensive change to an organism's genome than mutagenesis, which was widely used before genetic engineering was developed.

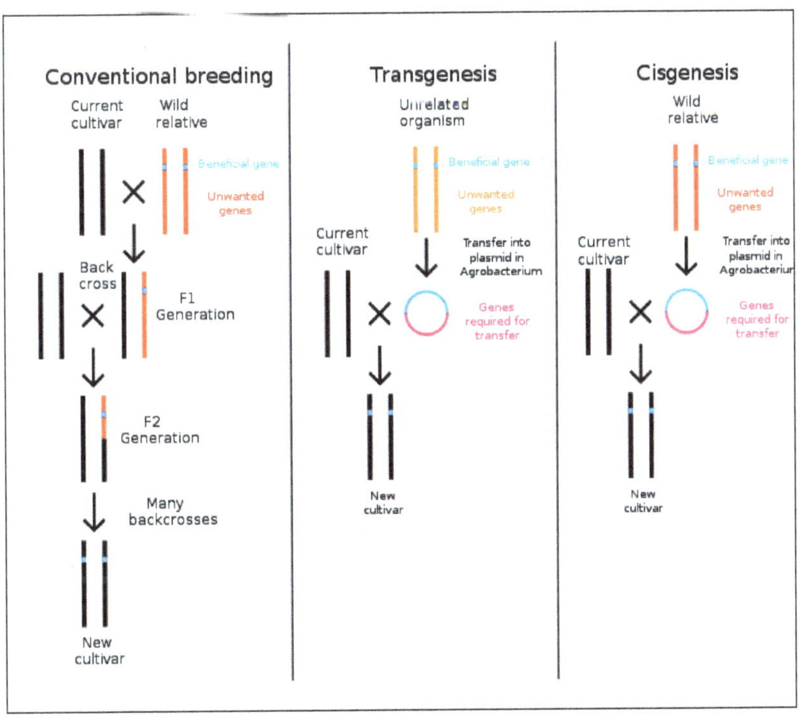

A diagram comparing the genetic changes achieved through conventional plant breeding, transgenesis and cisgenesis.

Some people believe that cisgenesis should not face as much regulatory oversight as genetic modification created through transgenesis as it is possible, if not practical, to transfer alleles among closely related species even by traditional crossing. The primary biological advantage of cisgenesis is that it does not disrupt favorable heterozygous states, particularly in asexually propagated crops such as potato, which do not breed true to seed. One application of cisgenesis is to create blight resistant potato plants by transferring known resistance loci wild genotypes into modern, high yielding varieties.

The Dutch government has proposed to exclude cisgenic plants from the European GMO Regulation, in view of the safety of cisgenic plants compared to classically bred plants, and their contribution to durable food production.

Agroinfiltration

Agroinfiltration is a method used in plant biology and especially lately in plant biotechnology to induce transient expression of genes in a plant, or isolated leaves from a plant, or even in cultures of plant cells, in order to produce a desired protein. In the method a suspension of Agrobacterium tumefaciens is introduced into a plant leaf by direct injection or by vacuum infiltration, or brought into association with plant cells immobilised on a porous support (plant cell packs), whereafter the bacteria transfer the desired gene into the plant cells via transfer of T-DNA. The main benefit of agroinfiltration when compared to the more traditional plant transformation is speed and convenience, although yields of the recombinant protein are generally also higher and more consistent.

The first step is to introduce a gene of interest to a strain of Agrobacterium tumefaciens. Subsequently, the strain is grown in a liquid culture and the resulting bacteria are washed and suspended into a suitable buffer solution. For injection, this solution is then placed in a syringe (without a needle). The tip of the syringe is pressed against the underside of a leaf while simultaneously applying gentle counterpressure to the other side of the leaf. The Agrobacterium suspension is then injected into the airspaces inside the leaf through stomata, or sometimes through a tiny incision made to the underside of the leaf.

Vacuum infiltration is another way to introduce Agrobacterium deep into plant tissue. In this procedure, leaf disks, leaves, or whole plants are submerged in a beaker containing the solution, and the beaker is placed in a vacuum chamber. The vacuum is then applied, forcing air out of the intercellular spaces within the leaves via the stomata. When the vacuum is released, the pressure difference forces the "Agrobacterium" suspension into the leaves through the stomata into the mesophyll tissue. This can result in nearly all of the cells in any given leaf being in contact with the bacteria.

Once inside the leaf the Agrobacterium remains in the intercellular space and transfers the gene of interest as part of the Ti plasmid-derived T-DNA in high copy numbers into the plant cells. The gene transfer occurs when the plant signals are induced and physical contact is made between the plant cells and bacterium. The bacteria creates a mechanism that burrows a hole and transfers the new T-DNA strand into the plant cell. The T-DNA moves into the nucleus of the plant and begins to integrate into the plants' chromosome. The gene is then transiently expressed through RNA synthesis from appropriate promoter sequences in all transfected cells (no selection for stable integration is performed). The plant can be monitored for a possible effect in the phenotype, subjected to experimental conditions or harvested and used for purification of the protein of interest. Many plant species can be processed using this method, but the most common ones are Nicotiana benthamiana and less often, Nicotiana tabacum.

Transient expression in cultured plant cell packs is a new procedure, recently patented by the Fraunhofer Institute IVV, Germany. For this technique, suspension cultured cells of tobacco (e.g. NT1 or BY2 cell lines of Nicotiana tabacum) are immobilised by filtration onto a porous support to form a well-aerated cell pack, then incubated with recombinant Agrobacterium for a time to allow T-DNA transfer, before refiltration to remove excess bacteria and liquid. Incubation of the cell pack in a humid environment for time periods up to several days allows transient expression of protein. Secreted proteins can be washed out of the cell pack by application of buffer and further filtration.

Silencing Suppressors in Agroinfiltration

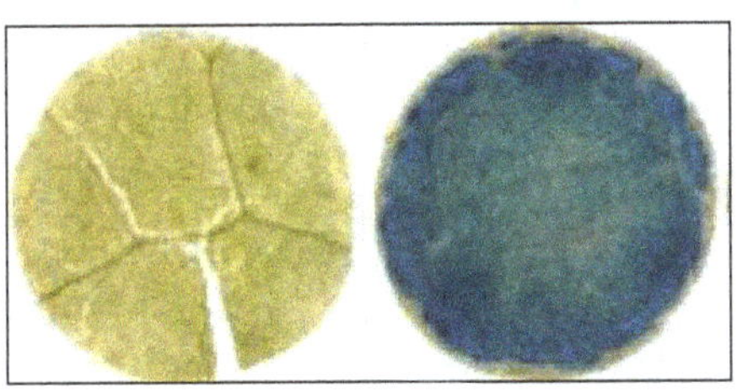

Agroinfiltration using a promoter: GUS construct in Nicotiana benthamiana" with TBSV p19 (right leaf disc) and without TBSV p19 (left leaf disc).

It's quite common to coinfiltrate the Agrobacterium carrying the construct of interest together with another Agrobacterium carrying a silencing suppressor protein gene such as the one encoding the p19 protein from the plant pathogenic Tomato bushy stunt virus (TBSV), or the NSs protein from Tomato spotted wilt virus (TSWV). TBSV was first discovered in 1935 in tomatoes and results in plants with stunted growth and deformed fruits. TSWV was discovered in tomatoes in Australia in 1915, and for many years was the only member of what is now known as genus Tospovirus, family Bunyaviridae.

In order to defend itself against viruses and other pathogens that introduce foreign nucleic acids into their cells, plants have developed a system of post-transcriptional gene silencing (PTGS) where small interfering RNAs are produced from double-stranded RNA in order to create a sequence specific degradation pathway that efficiently silence non-native genes. Many plant viruses have developed mechanisms that counter the plants PTGS-systems by evolving proteins, such as p19 and NSs, that interfere with the PTGS-pathway at different levels.

Although it is not clear exactly how p19 works to suppress RNA silencing, studies have shown that transiently expressed proteins in Nicotiana benthamiana leaves have an up to 50-fold higher yield when coinfiltrated with TBSV p19.

TSWV and other tospovirus NSs proteins have been shown to be effective as suppressors of both local and systemic silencing, and may be a useful alternative to p19 where the latter has been shown not to be effective. In other studies, p19 from Artichoke mottled crinkle virus has been shown to have a similar, although weaker, effect to TBSV p19.

Nuclease-based Techniques

Site-direct Nucleases Strategies for target and specific modification of the genome, or the genetic information, have been referred to as genome editing. Most genome editing techniques direct the repair of DNA doublestrand-breaks (DSBs) achieved in specific genome sequences by sequence-specific nucleases, or sitespecific nucleases, also called site-direct nucleases (SDNs). Such enzymes recognize and cleave genomic regions that shares sequence homology to their DNA binding enzymatic domains. The repair of DSBs can be directed to create a variety of targeted DNA sequence modifications, ranging from DNA deletions to the insertion of large arrays of transgenes. There are currently four major classes of SDNs: Meganucleases, zinc finger nucleases (ZFNs), transcription activator-like effector nucleases (TALENs), and clustered regularly interspersed short palindromic repeats (CRISPR)/Cas9 reagents.

These kinds of site-directed nuclease share two main features: (i) the capacity of binding to DNA at a specific sequence and (ii) to cleave those. The double stranded DNA break is then restored by the cell innate DNA repair machinery. DSBs stimulate the homologous recombination (HR) or error-prone non-homologous end-joining (NHEJ) DNA repair machinery, which can then lead to sitespecific mutagenesis or the integration of foreign DNA.

Site-directed nucleases are composed of two main parts: (i) the DNA-binding domain, which determines the DNA (site) specificity of the endonuclease; and the (ii) the DNA cleavage domain, which is responsible for the actual cleaving of the endogenous DNA (non-specific nucleases). Table summarizes the main components and modes-of-action of the four types of site-directed nucleases that will be further discussed later. And Figure is a schematic drawing of each nuclease in order to show basic structural differences between them.

Table: Overview of site-directed nuclease (SDN) types, their main components and modes of action. Note that 'bp' stands for base pair.

SDN type	Components	Mode-of-action
Meganucleases	Homing endonuclease with a large DNA recognition site (12–40 bp).	The catalytic domain also determines the DNA sequence specificity. Induces DSB in target DNA sequence.
Zinc-finger nucleases (ZFNs)	Multiple zinc finger peptides (each targeting 3 bp of genomic sequence) fused to restriction nuclease (usually Fok1).	ZFN targets are bound by two zinc-finger DNA binding domains (dimers) separated by a 5–7-bp spacer sequence. Induces DSB in target DNA sequence and the FokI cleavage occurs within the spacer.
Transcription activator-like effector nucleases (TALENs)	Arrays of the TAL effector motifs (each targeting 1 bp of genomic sequence) fused to restriction nuclease (usually Fok1).	The two TALEN targets sites (dimers) are typically separated by a 15-20-bp spacer sequence. Induces DSB in target DNA sequence and the FokI cleavage occurs within the spacer.
Clustered regularly interspersed short palindromic repeats (CRISPR)/Cas9	20-nt crRNA fused to tracrRNA and Cas9 endonuclease that has two nuclease domains and each cleaves one DNA strands.	The CRISPR-based guide RNA (gRNA) recognizes target DNA sequences by base pairing. Cas9 creates DSB at DNA target sites with complementarity to the 5'-end of a gRNA.

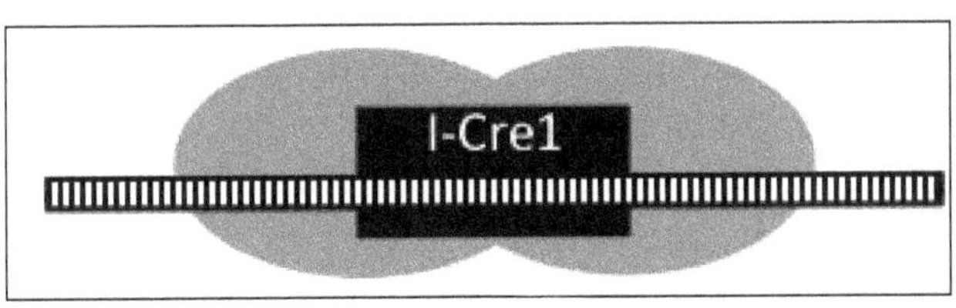

a. Meganucleases, specific I-Cre1 residues are responsible for target recognition and cleavage.

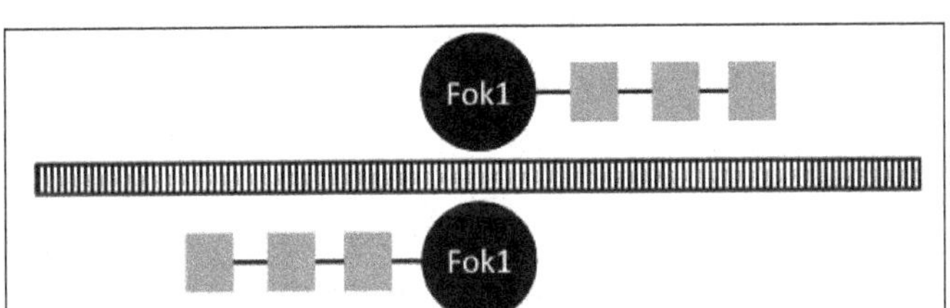

b. Zinc-finger nucleases, Fok1 is responsible for non-specific cleavage and target recognition is performed by the binding of two zinc-finger DNA binding domains (dimers) separated by a spacer sequence.

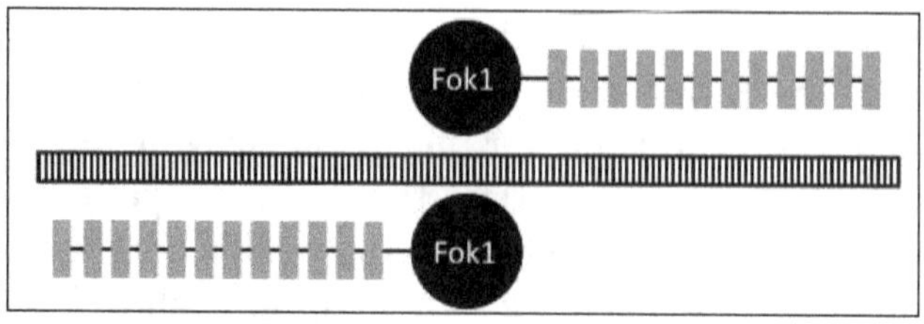

c. TALEN nucleases, two dimers also bind to DNA target sequence and these are fused to Fok1.

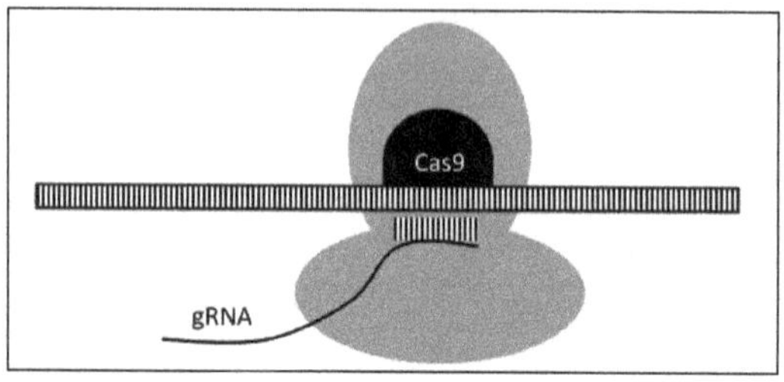

d. CRISPR/Cas9 system, DNA recognition is performed by a guide RNA strand and cleavage occurs by two nuclease domains and each cleave one strand of target DNA.

Types of Site-direct Nucleases and Modes-of-action

Meganucleases

Meganucleases were initially recognized as microbial endonucleases that catalyze genomic modification, rearrangement, protection, and repair in a gene-specific matter. The term meganuclease reflects the name "Omega" that was coined for the first known homing system as well as for designating sequence specific endonucleases proteins that could recognize large targets (from 12 to 40 bp). Homing is a process in which microbial self-splicing intervening sequences are specifically inherited, thus duplicated, into recipient alleles of their host gene that initially lacks the sequence. The term has progressively become synonymous with naturally occurring enzymes that bind and cleave large DNA sequence targets, and should no longer include artificial endonucleases such as zinc finger nucleases. This protein family has grown to include several hundred of identified members, found in eukaryotes, bacteria and archae, with the notable exception of metazoans and many could be shown to initiate homing events.

Over the past 25 years, a number of studies have applied meganucleases in animal and mammalian models for gene targeting recombination, thus showing the applicability of this technology Although several hundreds of other meganucleases had been identified in between the collection of cleavable se uences was too limited to address the comple ity of the genomes ues and Duchateau Engineering the sequence specificity of meganucleases was the next step to overcome this issue. But the lack of knowledge of protein structure and the relation between structure and function has not only delayed the development of new meganucleases sequences but also illustrated clearly the complexity of designing novel proteins.

The first homing endonuclease structure to be characterized was the I-Cre1 protein. Mutations of specific residues of I-Cre1 have shown substantial cleavage of new targets and loss of affinity to its natural site. Millions of combinations were screened resulting in several hundreds of mutants with locally altered specificity, but of course the randomization of the nine residues of the I-Cre1 meganuclease family (LAGLIDADG) would not result in enough variations to engineer the target site. In addition, most homing endonucleases are not clearly separated from the catalytic domains, thus making protein engineering procedures more complex and often compromising nuclease activity. This has hindered their widespread use.

To the best of our knowledge, there are only one publicly interest on the commercialization of a crop developed by the use of meganucleases. This was seen in a publicly available letter of inquiry to US Department of Agriculture (USDA) Animal and Plant Health Inspection Service (APHIS) made on the regulatory status of a product that had used meganucleases. Although little information was available, the inquiry in 2011 from a company called "Cellectis", required information regarding the regulatory status of a potential crop genome modification product derived from the use of I-Cre1 meganuclease. The I-Cre1 endonucleases are derived from the chloroplasts of Chlamydomonas reinhardtii and experimental selection and computational design methods have been further developed to re-engineer the cleavage specificity of I-Cre1 and its homologs.

Numerous methods have been developed to assess mutated meganucleases for activity and altered target specificities. The first report of an reengineered meganuclease to target a chromosomal locus of choice in a crop species, applied an reengineered I-Cre1 to target a sequence neighboring the maize liguleless locus and effectively generated heritable monoallelic and biallelic mutations at the target site, thereby knocking out the gene transcription.

These and other studies have shown that the modification of a homing endonuclease's cleavage specificity would be required in order to target and modify endogenous target sites in various biological genomes. Thus, a variety of biotechnology applications can potentially make use of the properties of a homing endonuclease that could lead to mutation, knockout, modification, or insertion of exogenous coding DNA into the endogenous gene target.

Zinc Finger Nucleases

Zinc fingers are small protein domains that are fold stabilized by a zinc metal molecule. These enzymes are structurally diverse and present among proteins that perform a broad range of functions in various cellular processes, such as replication and repair, transcription and translation, metabolism and signaling, cell proliferation and apoptosis. They work as interaction modules and bind to several compounds, ranging from nucleic acids to proteins and other small molecules.

Zinc fingers nucleases have emerged in the first attempt to stimulate gene targeting by applying novel chimeric restriction enzymes, fusing the catalytic domain of a well-known restriction enzyme, Fok1 from Flavobacterium okeanokoites, to another DNA binding domain to alter the DNA specificity of such enzymes. These researchers have enhanced gene targeting with designed zinc finger nucleases.

The ZFN architecture links the DNA-binding domain of a versatile class of eukaryotic transcription factors, so called the zinc finger proteins, with the nuclease domain of the Fok1 restriction enzyme. Originally, the number and order of cysteine and histidine amino acid residues was used to classify different types of zinc fingers (e.g., Cys_2His_2, Cys_4, and Cys_6). More recently, a systematic method classifies zinc finger proteins based on the overall shape of the protein backbone in the folded domain and the Cys_2–His_2 zinc-finger domain is among the most common types of DNA-binding motifs found in eukaryotes.

The flexibility of zinc finger proteins makes it a robust platform for the design of novel DNA binding domains. These domains contain a tandem array of Cys2-His2 fingers, each recognizing approximately 3 bp of DNA. In primary studies, three fingers were applied to bind a 9-bp target, which enabled ZFN dimers to specify 18 bp of DNA per cleavage site. But recent studies have added more fingers (up to six per ZFN) to specify longer and rarer cleavage targets. ZFN works as dimers because Fok1 must dimerize in order to cleave DNA. Thus, cleavage by Fok1 as part of a ZFN requires two adjacent and independent binding events, which must occur in both the correct orientation and with appropriate spacing to permit dimer formation.

The initial successes in plant genome engineering with ZFNs were reported in Arabidopsis thaliana and tobacco (Nicotiana tabacum). Further developments included a small number of crop plants, such as maize and soybean, and introduced important agronomic traits. Herbicide-resistance mutations were introduced into tobacco endogenous acetolactate synthase genes (ALS SuRA and SuRB) loci by ZFN-mediated gene targeting at frequencies exceeding 2% of transformed cells for mutations as far as 1.3 kilobases from the ZFN cleavage site. In addition, gene disruption and repair by the NHEJ pathway was observed in the tobacco protoplasts with more than 40% of recombinant plants having modifications in multiple SuR alleles. These authors have also tested for HR-based gene replacement by electroporation of ZFN plasmids into tobacco protoplasts along with SurB donor templates modified at positions known to confer herbicide resistance. Successful gene replacement was observed at frequencies ranging from 0.2 to 4%. Another example of gene editing using ZFNs was published by Shukla in which the authors have generated ZFNs that targeted the maize inositol-1,3,4,5,6-pentakisphosphate enzyme gene (IPK1). The disruption of IPK1 gene reduces the level of phytate, a known anti-nutritional compound found in seeds of many cereals. The IPK1 locus was disrupted by ZFN-mediated DSBs and repaired by HR using donor templates with an herbicide resistance gene and short, locus-specific homology arms.

Transcription Activator-like Effector Nucleases

TALENs are similar to ZFNs in that they have a DNA-binding domain derived from TALE proteins fused to Fok1 cleavage domain. TALE proteins are transcription factors from the plant bacterial pathogen Xanthomonas. The bacteria produce such proteins to target and control the expression of specific plant gene promoters in order to mimic host transcription factors and thus achieve successful pathogen infection.

DNA recognition by TALENs is achieved through arrays of TAL effector motifs. TALEs typically consists of a 16 to 20 single repeat monomers, with each monomer 34 amino acid residues in length. Although highly conserved, the exception is the hypervariable amino acid residues at positions 12 and 13, called repeat-variable di-residues (RVDs). The DNA-binding specificity of TALE repeats is apparently solely based on the nature of the RVDs in a typical 34-aa repeat, thus suggesting that the amino acids at position 12 and 13 probably interact directly with the DNA bases.

The consecutive array of repeats binds to a consecutive DNA sequence and rearranging repeat units generates novel custom-designed DNA-binding specificities with high potential for biotechnology. Computational and molecular biological analyses made it possible to decipher the TALE code for DNA recognition. Likelywise, the number of repeats and the sequence of the RVD region will determine the length and sequence of the target DNA. DNA-binding domains of individual TALENs typically contain between 15 and 30 RVDs and can then target 15 up to 30 nucleotides in the host genome accordingly. However, in contrast to zinc-finger proteins, the re-engineering of the linkage between repeats is not necessary to construct long arrays of TALEs because of the lack of interaction between every 3 nucleotides.

To date TALENs have been used to generate targeted modifications in a variety of plant species, such as Arabidopsis thaliana, tobacco and rice. Li have presented one of the first successful demonstrations of a nuclease-mediated modification of agronomic importance. These authors have exploited TALEN technology to edit a specific susceptibility gene Os11N3 (also called OsSWEET14) in rice to thwart the virulence strategy of Xanthomonas oryzae and thereby engineer heritable genome modifications for resistance to bacterial blight, a devastating disease for rice. During infection the bacteria secretes effector proteins that target DNA sequences in the promoter region of the rice sucrose-efflux transporter gene (OsSWEET14). It is interesting though, is that OsSWEET14 also plays a crucial role in the development of the plant, thus obtaining a knockout mutant to circumvent the effects of the pathogen was not feasible The authors successfully reduced the pathogen's virulence by disrupting the promoter sequence bound by the pathogen effector.

Clustered Regularly Interspersed Short Palindromic Repeats and Associated Proteins

Over the past two years, a more versatile genome editing tool has been developed based upon the bacterial clustered regularly interspaced short palindromic repeats (CRISPR) associated protein (Cas) adaptive immune system and tested in different organisms including plants. The first description of a CRISPR array was made in 1987 by researchers who found repeat sequences interspersed by short non-repeating spacer sequences in Escherichia coli, and in subsequent years, similar CRISPR arrays were found in Mycobacterium tuberculosis and other bacteria and archaea.

Bacterial CRISPR/Cas systems process short segments of foreign DNA into small elements that are then inserted into the CRISPR locus in the bacterial genome. RNAs derived from CRISPR loci are constitutively transcribed and processed into small RNAs (crRNA) of exogenously derived and incorporated DNA. After processing, these small RNAs guide other Cas proteins to mediate sequencespecific degradation of the foreign DNA thereby recognizing sequences encountered before in a immune-system like manner. The key components of the CRISPR/Cas system are Cas proteins (cleavage site) and CRISPR RNAs (recognition site). These two components are the basic constituents of this genome editing system and they occur under three major CRISPR/Cas systems.

Type I and III systems share some overarching features. In both, Cas endonucleases process the precursor CRISPR RNA (pre-crRNA) to crRNA that then assembles into a large multi-Cas protein complex capable of recognizing and cleaving nucleic acids with complementarity to the crRNA. In contrast, type II systems require a trans-activating crRNA (tracrRNA), with complementary sequences, to trigger pre-crRNAs processing by the double-stranded RNA-specific ribonuclease RNase III in the presence of the Cas9 protein. Cas9 is thought to be the sole protein responsible for crRNA-guided silencing of foreign DNA.

The type II system are considered simpler than other CRISPR systems, and its endonuclease family can be programmed with single RNA molecules to cleave specific DNA sites, thereby raising the exciting possibility of developing a simple and versatile RNA-directed system to generate DNA DSB for genome targeting and editing.

That is the major reason why the Cas9 system (CRISPR/Cas9) is the most explored Cas protein to date. It has been demonstrated that the two acting RNA molecules (crRNA and the tracrRNA) can be fused artificially to form a chimaeric RNA molecule termed single guide RNA (sgRNA). The site-specific catalytic action of this sgRNA–Cas9 complex is defined by a sequence of only ~20 consecutive nucleotides. It is reasonable that designing and producing a synthetic 20 nt sgRNA long is much simpler compared with the production of custom ZFNs and TALENs enzymes.

By modifying these basic constituents for the use in other organisms, the CRISPR/Cas9 system has been shown to be a useful tool for gene editing and silencing. CRISPR sequences can be engineered which give rise to crRNA directed against specific endogenous genes in various organisms. Significant advances for the use of this techni ue have been promoted by the generation of a "single-guide RNA" sgRNA that combines the function of the tracrRNA and crRNA in a chimeric molecule.

Kumar and Jain have recently reviewed the literature on the CRISPR/Cas system advances and opportunities for plant genome editing. These authors report that the usage of CRISPR/Cas9 systemmediated plant genome engineering are limited so far and it was only in 2013, that three different research groups simultaneously reported targeted genome modifications in plants using CRISPR/Cas9. Later, a number of other reports demonstrated the efficacy/potential of the CRISPR/Cas9 technology as an efficient genome editing system in model and crop plants.

An example of genome editing of crop plants was performed by Shan who designed several sgRNAs and codon-optimized Cas9, which target different DNA strands of four rice endogenous genes and one wheat gene. These authors have demonstrated different levels of mutagenesis (frequency of insertions and deletions) and the possibility of homology-directed repair by co-transformation of Cas9, sgRNA and single-stranded DNA oligos into plant cells. Another report demonstrated the use of a multiplex CRISPR/Cas9 technique for gene simultaneously edition of two sites (CHLOROPHYLL A OXYGENASE1 and LAZY1) in the Arabidopsis genome by applying a single CRISPR/Cas9 construct harboring two sgRNA expression cassettes. In addition, it has been demonstrated that the CRISPR/Cas9 system can provide homozygous gene editing within one reproductive generation.

Intended Modifications by Site-direct Nucleases

The outcome product of genome editing depends on mechanism used for the DSB repair and on the presence of a DNA template (donor DNA) that may be added artificially. Therefore, depending on the type of SDN and the presence of a template or not, the plant cell might trigger different repair mechanisms.

In line with the EFSA Panel on Genetically Modified Organisms Scientific opinion addressing the safety assessment of plants developed using Zinc Finger Nuclease 3 and other Site-Directed Nucleases with similar function and the Joint Research Centre Report on New Plant Breeding Techniques: State of the art and prospects for commercial development, the classification used for addressing such techniques was based on the types of modifications made by nucleases. In this regard, the intentional modifications should be categorized into three major classes:

SDN-1: SDN-1 techniques use SDNs to generate site-specific random mutations at target sequences harboring the nuclease recognition sequence. SDN-1 nucleases are delivered into plant cells without the presence of a donor DNA repair template. In most cases, single DSBs are repaired mainly by NHEJ. There are two kinds of outcome products of such class of SDN techniques. The first kind is resulted from unfaithful repair of NHEJ mechanisms that create small nucleotide deletions and insertions (i.e. indels). The second kind of outcome is the deletion of entire regions (e.g. e.g. to delete regulatory regions, exons/introns, whole genes or even parts of chromosome) between the two target sites when two DSBs are introduced. There is also evidence of plant gene targeting using SDN-1 techniques that have caused genomic inversions or translocations between the two induced DSBs.

SDN-2: SDN-2 techniques deliver SDN complexes along with an externally supplied DNA template (donor DNA). Donor DNA sequences contain the targeted desired mutations (i.e. few nucleotide substitutions or short insertion/deletions) and sequences with high homology to the endogenous gene target in order to trigger homologous recombination (HR). A successful HR event substitutes the endogenous sequence by the donor DNA introducing the desired mutations. The introduced modifications can be target to repair undesirable spontaneous mutations (i.e. targeted gene correction) or to introduce targeted mutations and new phenotypes (i.e. gene/allele replacement).

SDN-3: Similarly to SDN-2, SDN-3 techniques apply a SDN complex together with donor DNA between flanking DNA sequences showing homology to the endogenous target locus. However, in the case of SDN-3, a large stretch of donor DNA is used (i.e. plasmid DNA with promoter, gene of interest, terminator, gene reporter sequences, etc). Usually, homologous sequences will trigger HR, thus facilitating the target integration of exogenous sequences, which otherwise would integrate randomly in the genome. Therefore, the SDN-3 approach can be used for targeted addition of genes of interest, gene replacement and trait stacking.

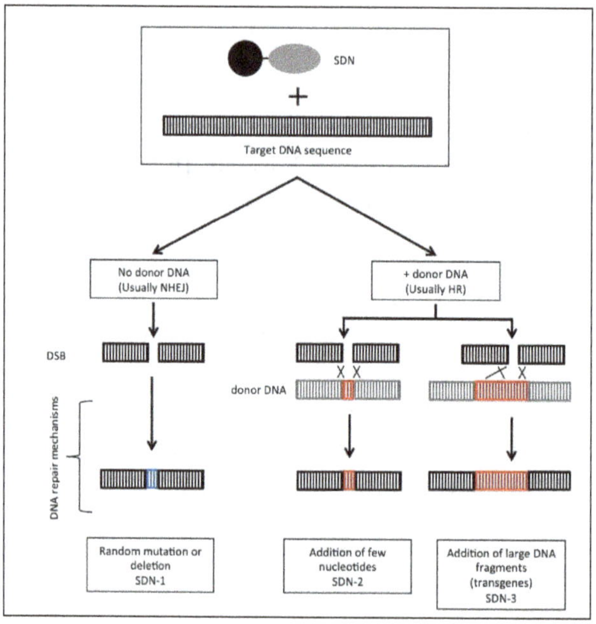

Schematic representation of intended modification by site-direct nuclease (SDN) types. Each SDN technique type (SDN-1, SDN-2 and SDN-3) is defined by the product outcome and will depend on the SDN design. This graphic shows a SDN in association with the target sequence. Double-stranded break (DSB) repair can occur via non-homologous end-joining (NHEJ) or homologous recombination (HR) when a donor DNA is present. SDN-1 can result in site-specific random mutations or deletion. SDN-2 can result in the addition of a few specific nucleotides. SDN-3 can result in the integration of large DNA fragments (transgenes).

Identification of Unintended Effects and Consideration of Uncertainty of SDN-based Crops

Site-direct nuclease-based techniques are subject to a number of possible unintended effects as any other technology applied to living organisms. In addition, due to the recent discovery of these new plant breeding techniques and the current lack of knowledge on the details of the mechanisms, significant uncertainties might be associated to the assessment of unintended effects.

We propose a step-by-step approach in order to contextualize and identify unintended effects and associated uncertainties of nuclease techniques on each stage of a SDN-based crop development. In order for a crop to be developed, other techniques than the genome editing technique itself are necessary for the delivery of the SDN and the regeneration and multiplication of the plant. Although we focus on the safety issues related to the use of SDN only, other biotechnology techniques (such as transgenesis, plant cell tissue culture, etc.) may be applied and also need to be considered.

Specificity of Target Cleavage

Genome stability is normally maintained in the face of potentially dangerous DSBs by the combined activity of non-homologous DNA end-joining and homologous recombination. In gene targeting approaches, the location of inducible genomic DSBs will depend on the DNAbinding specificity of the nuclease. The specificity will then determine the activity and toxicity of nucleases.

Off-target cleavage has been most studied in ZFNs. Although at different levels, the imperfect specificity of engineered zinc-fingers domains has been linked to cellular toxicity.

Off-target cleavage of ZFNs has been investigated by Pattanayak and coworkers utilizing an in vitro selection method. The authors find that information about the specificity of ZFNs has been based on unproven assumptions that (i) dimeric ZFNs cleave DNA with the same sequence specificity with which isolated monomeric ZF domains bind DNA, and (ii) the binding of one ZF domain does not influence the binding of the other ZF domain in the functional dimeric protein. As a result, DNAbinding specificities of monomeric ZFN domains have been used to predict potential off-target cleavage sites of dimeric ZFNs. The methodology proposed by Pattanayak identified several sites in the human genome that can be cleaved in vitro by two different ZFNs, they have then further examined these sites for evidence of ZFN-induced mutagenesis in cultured human cells expressing both ZFNs. The authors have confirmed that cleavage was likely to be dependent on cell type and ZFN concentration. Doyon have also reported that transient hypothermia increased ZFN-mediated gene disruption in several cell lines between 1.5–15-fold. Although it is not clear yet whether this observation is generally applicable to any system, the authors showed that the environment (i.e. temperature shift), can influence ZFN protein levels in the cells and thus have an impact on nuclease activity. Similar results were also obtained by Pruett-Miller who showed that by regulating protein levels, they could maintain high rates of ZFN mediated gene targeting was maintained while ZFN toxicity was reduced.

In regards to cell-type off-target cleavage dependency, other authors have also observed that ZFNs that can cleave at sites in one cell line may not necessarily function in a different cell line most likely because of local differences in chromatin structure. In addition, their results indicate that excess DNA-binding energy results in increased off-target ZFN cleavage activity and suggest that ZFN specificity can be enhanced by (i) designing ZFNs with decreased binding affinity, (ii) by lowering ZFN expression and (iii) by choosing target sites that differ by at least three base pairs from their closest sequence relatives in the genome. Furthermore, the authors show that a fourfinger ZFN showed a more diffuse range of cleavage positions with relaxed specificity and a higher tolerance of mutant sequences than a three-finger ZFN. Nonetheless, half or more of all potential substrates with one or two site mutations could be cleaved by ZFNs, suggesting that binding affinity between ZFN and DNA substrate is sufficiently high for cleavage to occur even with suboptimal molecular interactions at mutant positions. That same study also reveals that in vitro spacer preferences do not necessarily reflect spacer preferences in cells. Their results suggest that the dimeric FokI cleavage domain can influence ZFN target-site recognition.

Consistent with these results, Gabriel have investigated the specificity of ZFN by a different method that was based on detecting off-target DSBs that have captured integrase-defective lentiviral vectors by NHEJ. These authors have analyzed genome-wide integration sites mapping in vivo cleavage activity of four ZFN pairs. Their results showed that the ranking of in vivo cleavage sites could not be predicted in silico. Thus, indicating the need for comprehensive maps of ZFN activity in in vivo models.

Other groups have also put their efforts in alternative strategies to decrease ZFN off-target cleavage. Doyon for example, have identified critical residues involved in off-target homo-dimerization activity of ZFN and they have successfully designed ZFN suppressing this homo-dimerization. An extensive review about the parameters (i.e. DNA-binding domain, DNA-cleavage domain, Inter-domain linker and others) that improve ZFN activity and specificity is provided by ndel and athomen These authors highlight the need for a proper assessment of each type of ZFN toxicity in order to be able to distinguish between cytotoxicity, genotoxicity, immunotoxicity or teratogenicity.

Undesired off-target cleavage has been also recognized for TALENs. According to several research groups, this remains an important issue and may cause severe side effects to the model organism under investigation.

Similar results to ZFNs were obtained when assessing TALEN specificity. Guilinger have profiled 30 unique TALENs with varying target sites. Computational analysis predicted 76 off-target substrates in the human genome, 16 of which were accessible and modified by TALENs in human cells. In addition, their results suggested that (i) TALE repeats bind DNA relatively independently; (ii) longer TALENs are more tolerant of mismatches, yet are more specific in a genomic context; and (iii) excessive DNA-binding energy can lead to reduced TALEN specificity in cells. TALEN protein concentration has been also observed as a factor contributing to off-target cleavage. Tesson have observed results suggesting dose-dependent mutation rates. TALE arrays are reported with on-target DNA binding affinities sufficient enough to theoretically saturate target sites even when expressed at modest, mid-nM concentrations in the cell.

Fine have recently highlighted that experimental identification of ZFN and TALEN offtarget sites is a daunting task because of the availability and size of host genomes and the large number of potential cleavage sites to assay. More importantly, these authors also present literature from previous attempts to identify new off-target sites that were based entirely on bioinformatics search methods and that have all failed to locate any off-target cleavage sites, which has led to the belief that identifying off-target activity based on sequence homology alone would not be fruitful. Several time-consuming experimental approaches were then developed, such as SELEX, bacterial one-hybrid, in vitro cleavage or IDLV trapping. But the overall conclusion of the recent literature review is that although several experimental and theoretical methods have improved ZFN and TALEN activity, genome-wide specificity of ZFNs is still under scrutiny.

The historical record about the determinants of the specificity of CRISPR/Cas9-based techniques seems even less substantial than those for ZFNs and TALENs. Hallmark papers on comprehensive analysis of off-target effects of CRISPR/Cas-derived RNA-guided endonucleases and nickases date to a year ago or even later. Very recently, three groups independently showed that RNA-guided nucleases (RGNs) indeed induce off-target mutations, even at sites that differ by 5 nt from target sites in human cells. Fu have applied a human cell–based reporter assay to characterize off-target cleavage of CRISPR-associated Cas9-based RGNs. These authors found that off-target sites were mutagenized with frequencies comparable to (or higher than) those observed at the intended on-target site. In addition, their work demonstrates that RGNs can be highly active even with imperfectly matched RNA-DNA interfaces in human cells, a finding that might confound their use in research and other therapeutic applications.

Behaj have reviewed targeted mutagenesis in model and crop plants that have used CRISPR/Cas system. These authors show the outcomes of four reports that have addressed CRISPR system to off-target activity when applied in plants. While Nekrasov did not detect offtarget activity in N. benthamiana for 18 off-sites with sequence similarity to the target; two of the four reports detected experimental evidence of off-target activity in rice.

These reports have shown that the 3' end of the guide sgRNA sequence is the one responsible for target specificity of the CRISPR/Cas system. In addition, researchers have also identified a protospacer adjacent motif (PAM) 20 bp right after the target sequence as essential for Cas9 cleavage activity. Mismatches located within the last 8-10 bp of the 20 bp target sequence often block target recognition by as9 while mismatches towards the 5' end of the target are better tolerated. Although this might work as a general rule, Hsu reported that a noncanonical variant sequence of the PAM retains some activity and that the number and position of acceptable mismatches is target-dependent and should not be generalize the reported rates. Most importantly, Behaj highlight that the overall number of tested off-sites in all CRISPR/Cas studies was relatively small and general conclusions would be premature. These authors strongly suggested comprehensive studies based on whole genome sequencing of mutant plants in order to fully address the off-target cleavage issue in planta.

The recent literature demonstrates off-target cleavage activity of all nuclease types under different various experimental conditions. In addition, it is clear that there are recognized uncertainties regarding the probabilities and the characteristics by which these off-targets are generated. The difficulty in predicting these effects leads to the necessity of prior in vivo testing of these nucleases activities, as well as comprehensive untargeted profiling approaches (such as omics techniques) in order to detect these off-target cleavage sites in the host genome. The inheritance of such unintentional genomic changes is even less studied and cannot be assumed to vanish from the host genome organism just by crossing.

Specificity of Double-stranded Break Repair Mechanisms

Considering that DSBs can occur at target and off-target sites, it is necessary to verify intentional and unintentional mutations that might have resulted from the cell innate DNA repair mechanism at all DSB sites. DSBs can be repaired via two main pathways: homologous recombination (HR) and non-homologous end-joining (NHEJ).

HR requires regions of DNA homology and repairs DSBs by using information on the undamaged sister chromatid or, in the case of genome editing techniques, the donor DNA. Whereas sequences for HR are linked via regions that are identical to one another, sequence information does not play a major role in the rejoining of the two DSB ends for NHEJ. Therefore, NHEJ uses extremely limited or no sequence homology to rejoin juxtaposed ends in a manner that is not required to be error free.

The question whether the application of a SDN into the plant cell will trigger HR or NHEJ will depend on many factors, and yet most of them are still unknown. This is because all breaks are not equally created and the particular repair mechanism to be used will depend largely on what is possible and needed based on the structure of the broken DNA. In addition, it is important to mention that the frequency of SDN-mediated HR has been also shown to be dependent on the architecture of the donor DNA. Critical parameters range from the length of the homology arms and non-homologous sequences to the linear or circular DNA donor topology. The question to what extent the homology arm lengths as well as the topology and the configuration of the donor DNA influence the ratio between HR and illegitimate NHEJ has not been conclusively answered yet.

Consistent with these evidences, Ramirez highlights that although HR-mediated alterations can be efficiently introduced using engineered nucleases, alleles can also acquire NHEJmediated mutations; thus, unwanted alterations at other off-target genomic sites can also be introduced by NHEJ-mediated repair. And as an alternative to the potential undesirable consequences of introducing DSBs in living cells, these authors propose a single-stranded break model (SSB or nicks) to genome editing. They have converted engineered ZFNs into zinc finger nickases by inactivating the catalytic activity of one monomer in a ZFN dimer. The concept applying nicks for stimulation of HR has been previously suggested in the context of theoretical models and recombination induced by RAG proteins.

A very recent comparative analysis of off-target effects of CRISPR/Cas-derived RNA-guided endonucleases and nickases has shown improvements in nuclease specificity. These authors have measured mutation frequencies of paired nickases and nucleases at off-target sites using deep sequencing. Cas9 nucleases induced off-target mutations at six sites that differ by 1 or 2 nt from their corresponding on-target sites with frequencies that ranged from 0.5% to 10%. In contrast, paired Cas9 nickases did not produce indels above the detection limit of 0.1% at any of the six off-target sites.

These mutations can have a variety of effect on endogenous gene expression. Small deletions or insertions can cause gene knockout, or be completely ineffective from normal expression rate perspective among other impacts. Some of these mutations might not represent a phenotypic change and thus be detected by an unsupervised eye, but still trigger safety considerations Furthermore, it cannot be assumed that they will not be inherited by daughter cells. It is clear at this stage that knowledge about the DSB DNA repair mechanisms applied by host cells that have been exposed to nucleases is still in its infancy. Most interestingly, the large majority of the studies are still based on mammalian models and not plant models. The offtarget cleavage of such nucleases seems to be reaching a consensus in the scientific community but the outcomes of such breaks as well as the repair mechanisms are far from being elucidated.

Oligonucleotide-directed Mutagenesis-based Techniques

Oligonucleotide-directed mutagenesis (ODM) is an approach that uses oligonucleotide (25-200 nucleotides or base pairs (bp) in length) or polynucleotide (>200 nucleotides or bp in length) sequences to induce sequence-specific modifications (i.e. mutations) of endogenous genomic target sequences. Oligo/polynucleotide generally carry a limited number (between 1 and 5) of modifying bases within a segment of DNA that is otherwise homologous to the target sequences. Most importantly, this strategy does not use viruses as delivery vehicles; neither does it aim to add more genetic material to the target gene unless, otherwise combine to other genome editing or transgenesis techniques.

Moerschell, were the first to show that single-stranded DNA oligonucleotides could direct in vivo base changes in Saccharomyces cerevisiae. To date, numerous publications report that synthetic oligonucleotides can induce small sequence changes in nuclear DNA of eukaryotic cells. The genetic changes that can be obtained using oligonucleotide-directed mutagenesis include the introduction of a new mutation (replacement of one or a few base pairs), the reversal of an existing mutation (gene repair) or the induction of short deletions in native genomic sequences.

The general mechanism involved in oligonucleotide-directed mutagenesis-based gene repair is believe encompass a two-step process, namely (i) DNA strand pairing and (ii) DNA repair (also called gene conversion). However, the exact mechanisms by which each of these pathways occur seems controversial in the literature. There are two main models that explain the results obtained by several research groups. This report does not aim at deciphering the details of each proposed model, but to show discrepancy on understanding the modes-of-action by different research groups. This highlights knowledge gaps in the basic functioning of such technique and also shows that most assays are error-prone trials with large variation in the efficiency. In addition, most of these mechanistic studies are based on animal evidences and thus cannot be fully applied to plants.

The strand paring is the first step for ODM. Strand pairing involves the synthetic molecule that is brought into homologous region at the site in the chromosome that is designated for change, and the event is catalyzed by members of the RAD54 epistasis group in plants. The invasion and binding of oligonucleotides with the target sequence resulting in the formation of a loop structure often referred to as D-loop. Whether this step is a rate-limiting step in the process of gene correction by oligonucleotides is yet controversial.

In the second step, after alignment, one or few base mismatches are present between the oligonucleotide molecule and the target strands. The following steps are controversial in the literature. In addition, it has been also presented evidence that the different types of oligonucleotide molecules might trigger distinct cell responses. Here it is presented a summary of the two most accepted mechanistic models for the DNA repair pathways. Figure is a schematic representation and a summary of the main steps:

- In one model, Kmiec's group believes that several pathways can be triggered by the oligonucleotide molecule once it is paired. The molecule can be simply detached from the genome and no modification is observed or, alternatively, the oligonucleotide assimilates into the target site and replaces one of the parental strands, perhaps integrating directly into the helix, and the synthetic oligonucleotide strand is further removed from the parental target strand by nuclease activity. In a third scenario, the mismatched base pair (formed between the vector and the target) is corrected or reversed by DNA repair enzymes, which are probably members of the DNA mismatch-repair family. There is, however, a fourth explanation from other data that implicate nucleotide excision repair as an active pathway in gene repair.

- Yet another research group led by Igoucheva propose an alternative model for gene repair mechanism. In this model, an oligonucleotide is assimilated to the target DNA during active transcription, leading to formation of a transient D-loop structure. The trafficking of RNA polymerase is interrupted by the D-loop, and the stalled RNA polymerase complex may signal for recruitment of DNA repair proteins, including transcription-coupled DNA repair and nucleotide-excision repair.

But not only the mechanisms by which oligonucleotide-directed mutagenesis occur but also the nomenclature of such agents and approaches are not consensus, further complicating the literature search. Lusser , for instance, report countless descriptive and operational names to designate ODM, these are: oligonucleotide-mediated gene modification, targeted gene correction, targeted gene repair, RNA-mediated DNA modification, RNA-templated DNA repair, induced targeted mutagenesis, targeted nucleotide exchange, chimeraplasty, genoplasty, oligonucleotide- mediated gene editing, chimeric oligonucleotide- dependent mismatch repair, oligonucleotide-mediated gene repair, triplex-forming oligonucleotides induced recombination, oligodeoxynucleotide-directed gene modification, therapeutic nuclcic acid repair approach.

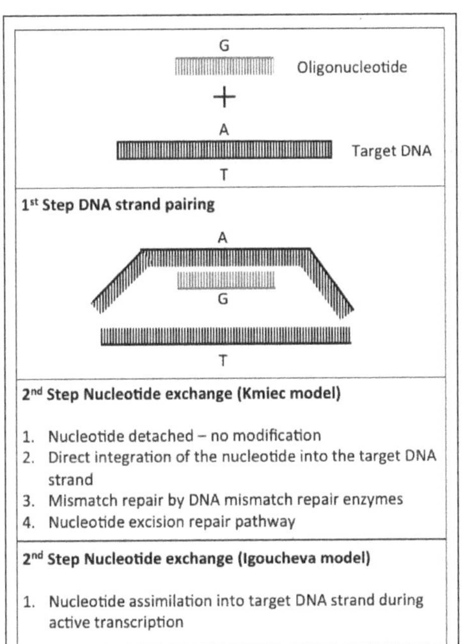

Schematic representation and summary points of the main theoretical models for oligonucleotidedirected mutagenesis (ODM) mode-of-action. The oligonucleotide is delivered into cells (first square), and it targets DNA strand, the process is known as the DNA pairing step (1st Step). The second step (DNA repair step) is explained by two main models. In Kmiec's model four main possibilities take place in the cell and are based on the activity of mismatch repair enzymes or the direct exchange (integration or excision) of mismatched nucleotides. In Igoucheva's model, mismatched nucleotides are exchange during transcription activity in a transient D-loop DNA structure.

At present, the potential applicability of gene repair by ODM seems to be limited by the lack of knowledge on how these molecules are functioning and, consequently, by low efficiency of correction. For most ODM specialized researchers, mechanistic studies to identify critical proteins involved in the gene correction process will undoubtedly play an important role for designing rational approaches toward increasing the targeting efficiency, efficacy and safety.

Types of Oligonucleotide-directed Mutagenesis

Different types of oligonucleotide-directed mutagenesis methods shows that different types of oligonucleotide molecules have been applied to various cell species with distinct cell-type versatility and apparent various efficiencies of gene modification.

In order to enhance the understanding about modes-of-action and implications of each described oligonucleotide type, we will follow the categorization provided by Igoucheva, Laible and Sargent:

<table>
<tr><th>Oligonucleotide type</th><th>Components</th><th>Mode-of-action*</th></tr>
<tr><td>Single-stranded oligodeoxynucleotides (SSOs or ssODMs)</td><td>ssDNA (<200-nt oligonucleotide)</td><td>DNA strand pairing and formation of a D-loop region that triggers DNA repair pathways.</td></tr>
<tr><td>Chimeric RNA-DNA oligonucleotide molecules (RDOs)</td><td>ssRNA-DNA or dsRNA-DNA oligonucleotide</td><td rowspan="3">Model 1: direct integration of base mismatches into the helix or corrected by DNA repair enzymes, probably mismatch-repair family. Or yet via nucleotide excision repair

Model 2: via transcriptional-coupled DNA repair enzymes, or nucleotide- excision repair systems or mismatch repair mediated by recombinases.</td></tr>
<tr><td>Small Fragment Homologous Replacement (SFHR)</td><td>dsDNA ou ssDNA (200–2000-nt polynucleotide)</td></tr>
<tr><td>Triple helix-forming oligonucleotides (TFOs)</td><td>ssDNA (10–40-nt oligonucleotide) that target poly-purine or polypyrimidine regions of DNA</td></tr>
</table>

Single-stranded Oligo-deoxynucleotides

SSOs are single stranded oligonucleotides molecules of about< 200 bp in length and comprised of a single mismatch to the target sequence that is generally in the middle of the molecule. Although SSOs have been synthesized with phosphorothioate backbone's -O-methyl uracil or with 50 or 30 thymidine clamps to inhibit degradation, they have also been used without modification to facilitate homologous exchange.

The first studies using homologous SSOs of 20-70 nucleotides containing a single mismatch were shown to trigger defined point mutations in mammalian cells and yeast. The frequency of correction ranged from 10^{-7} from 10^{-3} (one in 10th million cells to one in a thoursand cells) in mammalian cells and yeast, respectively, depending on the amount, length, and polarity of SSO, as well as the genetic background of the recipient yeast.

The activity of such small molecules seems to be influenced by several biological factors, thus leading to low efficiency rates. However, high correction rates of approximately 0.05% for nuclear extracts, 1% for episome, and 0.1% for chromosome of mammalian cells were observed by Igoucheva when applying relatively short single-stranded oligodeoxynucleotides, 25– 61 bases homologous to the target sequence except for a single mismatch to the targeted base in the mutant b-galactosidase gene. Igoucheva have made an extensive review of studies that utilize SSOs and concluded that a large variation in frequencies of gene alteration has been observed by researchers.

Accordingly, several studies on SSO do not report coherent or consistent results. While some studies suggest that transcription may be a factor in the SSO correction process due to a higher efficiency of correction with the anti-sense non-transcribed SSO target; others attribute this result to the context of chromatin accessibility due to the integration of ODM transgenes in different regions of the genome. According to Sargent these would represent chromosomal environment differences in terms of gene expression. On the other hand, Dekker also suggested that there are no strand-associated differences in SSO-mediated targeting.

The methodology seems to be currently limited by its low frequency of repair events, variability amongst different experimental settings, and low viability of mutated cells. But whether contradictory findings reflect differences in the cells used in the targeting studies requires further analysis.

Chimeric RNA-DNA Oligonucleotide Molecules

Chimeric RNA-DNA oligonucleotides are composed of a contiguous stretch of RNA and DNA residues has been developed to facilitate correction of single-base mutations in mammalian cells. The original design of the chimeric RNA-DNA oligonucleotide consisted of a double-hairpin capped duplex comprising a 25-nucleotide-long DNA stretch (DNA-strand) containing the mismatch, paired to a fully complementary RNA stretch (RNA-strand) with a pentameric DNA interruption in the middle.

While the first constructs were double-stranded oligonucleotides that consisted of a one DNA strand and another strand with a central DNA stretch but terminal regions of RNA, the second generation RDOs single-stranded DNA oligonucleotides with terminal RNA stretches or simply uracil residues. The idea of replacing the complementary DNA strand to a RNA strand emerged from the observation of previous studies that showed increased efficiency in RNA-DNA chimeric molecules.

Chimeric double-stranded oligonucleotides are designed to be self-complementary and produce double-hairpin structures. Within the duplex conformation, two complementary strands can be distinguished: a chimeric strand consisting of an interposed DNA fragment (5 bp) and two stretches of RNA (10 bp) flanking this region, and another strand that is composed only of DNA contains a mutation relative to the homologous target sequence.

The first experimental evidence of successful gene correction by RDO was demonstrated in 1996 by correction of a point mutation in alkaline phosphatase cDNA in Chinese hamster ovary (CHO) cells. The RNA-DNA sequence was designed to align with the sequence of the mutant alkaline phosphatase locus and to contain the desired nucleotide change. The chimeric molecule was introduced into human cells containing the mutant gene on an extra-chromosomal plasmid, correction of the point mutation was accomplished with a frequency approaching 30%. The design of this RNA-DNA hybrid molecule makes them highly active in homologous pairing reactions in vitro and the observations suggested that recombination of these exogenous RNA-DNA sequences is protected from exonucleolytic degradation by capping both ends.

Some experiments have extended the utility of RDO molecules to plants. Apparently, this seems to be the preferred oligonucleotide types used in plant studies. For example, Kmiec used a plant cell-free extracts system to study DNA-repair mutants and the role of plant proteins in the DNA repair process. In this assay, cell-free extracts were mixed with a plasmid containing a specific mutation (point or frameshift mutations).

Interesting to note, large variations among RDO experiments were observed and these have been attributed to the use of different species, cell types, genes, methods of delivery and assays, which have hindered efficiency comparisons.

Small Fragment Homologous Replacement

A widely used approach to generate transgenic animals was based on the delivery of homologous DNA and selectable markers to embryonic stem cells. In principle, a portion of a gene can be introduced into the nucleus allowing pairing of a DNA fragment with its homologous loci by means of proteins involved in HR, thus modifying genomic DNA in a sequence-specific manner. This approach has been further refined to the use of SFHRs for relatively high level of targeted correction (1%) and achieve therapeutic level.

SFHRs are distinct from SSO because they are larger molecules (generally between 200 and 2000 bp/nt in length) and can be individual ssDNA, complementary ssDNA, or double-stranded DNA. In addition, SFHR relies on two main features (i) a SFHR carrying a single or multiple base alterations, finding its sequence homolog, and (ii) the cellular enzymatic pathways that facilitate homologous exchange between target sequences and the SFHR. Although there has been speculation about the mechanism s that underlie SFHR, there is still only limited knowledge of the cellular factors that influence SFHR. To the best of our knowledge, there have been only a limited number of studies that have explored the mechanisms that underlie SFHRmediated modification. Even fewer apply SFHR in plants.

Triple Helix-forming Oligonucleotides

TFOs are single-stranded onligonucleotides, generally 10–40 nt in length, that bind to specific regions in duplex DNA as a third strand to form a triple helix at poly-purine or poly-pyrimidine regions of DNA bound via Hoogsteen hydrogen bonds.

The triple-helix-forming oligonucleotide recognizes the sequence surrounding a targeted base and the coupled reactive group or DNA elicits DNA repair and recombination process, resulting in sequence alteration. Such oligonucleotides have been used previously to change targeted DNA sequences by approximately 1% but are limited by the restrictions of the target sequence, which must consist of homopurine or homopyrimidine stretches for triplex formation.

TFOs have a bipartite structure that contains sequences that are favorable for triple helix formation and donor DNA sequences homologous to the chromosomal target sequence to be modified. Although covalent linkage of the TFO portion of the oligonucleotide was initially used to orient the contiguous DNA sequence to make available for repair of the DNA, most studies now use unlinked TFOs and donor oligo/polynucleotides to stimulate HR and other DNA repair pathways. In addition, TFOs coupled to functional chemical groups or double-stranded DNA have been also used to alter, specifically or non-specifically, genomic DNA in cultured cells and in vivo.

Studies evaluating the enzymatic pathways that may underlie the TFO-mediated homologous exchange indicate that nucleotide excision repair is involved. We have not found plant examples utilizing TFOs in the recent literature. Many examples can be found in mammalian and other animal models.

TFOs can also be utilized to create permanent heritable changes in the genome by both site-specific mutagenesis and site-specific recombination. However, mechanistic questions remain unanswered. The exact mechanisms behind cellular processes that play a role in mediating these genetic changes are unknown. There is a need to determine which factors can influence the availability of genomic target to DNA binding molecules, transcriptions levels, cell cycle phase, nucleosome binding and histone state that may all play important roles.

Identification of Unintended Effects and Consideration of Uncertainty of ODM-based Crops

The identification of similarities and differences in the mechanisms underlying the various ODM gene-targeting approaches has been hindered by various experimental factors, such as the variability in the exogenous DNA, the genetic target, the cell systems employed, and the approaches used to analyze the efficiency of homologous exchange.

Similar to SDN techniques, ODM-mediated homologous exchange process relies on the native cellular machinery to perform the exchange between the exogenous nucleotide sequences and the endogenous target sequences. And since there are multiple DNA repair and replication pathways that depend not only on the type of DNA modification, but also on the cell type and the stage of the cell, it is difficult to generalize about the involvement of specific mechanisms in the homologous exchange process for an individual strategy.

The situation becomes even more complicated when most of the ODM examples were performed in animal models, with a very few applications in plants. With that in mind, it is relevant to state that the discussion on unintended effects below are restricted to animal examples just because this has never been studied in plant species.

It is clear that the events of the second phase where the 'correction' of the targeted mutant gene actually occurs are poorly understood. The current challenge centers on the fate of the oligonucleotide paired at the target site and its role in the correction event. From Parekh-Olmedo perspective, the answers may lead to an understanding of not only how it works but also which cellular factors are involved in its regulation.

Olsen have tested a direct in vivo gene modification approach using a CHO cell assay containing a stable integrated mutated enhanced green fluorescent protein (EGFP) reporter. These authors observed specific correction of mutated EGFP to a functional wild-type sequence after oligonucleotide administration in vivo. Most interestingly, they have observed that the repair of sitespecific DS break by homologous recombination is severely reduced in daughter cells compared to the paternal V79-4 cell line. The authors further mention the fact that different chromosomal integration of the EGFP transgene in daughter and parental cell lines might have influenced these results. They also observed that both predicted and unpredicted alterations have occurred in the mEGFP2 template. In addition, 9.3% of the amplified sequences have a corrected stop codon but also show additional flanking mutations that are not contained in the oligonucleotide. The interesting work of Olsen have also shown high apoptosis rates that has been suggested to be coupled to the targeted sequence alteration rather than to general toxicity. But whether the apoptosis observed in the experiments is a general problem associated with the presence of large amounts of oligonucleotides in cells, whether it is a more specific process related to cells that have undergone sequence conversion and whether any of the above mechanisms account for the observed selective apoptosis should be further investigated.

Liu have also observed that during SSO-mediated gene repair reaction, a significant population of corrected cells failed to divide, and were much more prone to undergo apoptosis. In addition, their results showed that while a defective mismatch repair mutant greatly enhanced the efficiency of gene correction, compromising the mismatch repair system did not yield any viable corrected clone, indicating that the mismatch repair machinery, although plays a critical role in determining SSO-directed repair, was not involved in the observed cellular genotoxic responses.

On the other hand, Parekh-Olmedo highlights that such cellular apoptotic responses should not be surprising but rather predicted due to the common high oligonucleotide dosage delivery into cells. These authors also recall the fact that it is widely recognized that free ends induce a DNA damage response, which leads to cell cycle arrest, replication fork stalling and the activation of repair pathways, including HR. And that the combination of these activities determines the fate of the cell; whether the damage will be repaired or the cells will be programmed for death.

Not only the amount of oligonucleotide delivered into cells, but also length and polarity have shown to highly influence chromosomal gene corrections. The strand orientation, sense or anti-sense, showed a 1000-fold difference in frequency, indicating a possible influence of transcription on gene correction.

The large variability in the results obtained by these experiments shows the numerous factors that influence the functioning of ODM and the huge knowledge gaps surrounding them. In addition, some of these gaps have been followed by the application of wrong experimental methods. Igoucheva highlights that most studies have detected the gene conversion event by PCR amplification of genomic DNA, without direct confirmation at the protein level. Consequently, concerns have been raised about a potential artifact during PCR amplification or by cross contamination between cells.

Yoon suggest that the delivery and quality of an oligonucleotide, two favorite candidates for the reported variation among different systems, are not sufficient to contribute the large variation seen in experiments. These authors revealed that mechanistic studies from their group and others showed that the DNA-strand of oligonucleotide, for examples, might have strong influence in the gene repair activity. They further suggest that, based on these results, the active domain (DNA strand) needs to be exposed at the right time and at the right locale to the target DNA for successful gene correction. But since the processing of an oligonucleotide by nucleases or helicases to expose the active domain is random, the subsequent strand-pairing reaction to the target DNA is also random and hard to control.

There seems to be increasing evidence that links cellular toxicity to unintended effects from oligonucleotide delivery into cells. But a specific focus on efficacy of target correction. This might be explained by the uncertainties on the modes-of-action of such molecules. It seems that, unless more experimental evidence elucidates the mechanisms by which these molecules function, robust investigation about possible unintended effects will remain marginal.

References

- Plant-biotechnology-2, applications-of-rapd-marker-in-plants-genetics-72287: biology discussion.com, Retrieved 11 May, 2020
- Gulcher J (April 2012). "Microsatellite markers for linkage and association studies". Cold Spring Harbor Protocols. 2012 (4): 425–32. doi:10.1101/pdb.top068510. PMID 22474656
- Dna, dna-markers, dna-markers-definition-properties-and-applications-37972: biologydiscussion.com, Retrieved 29 June, 2020
- Jochemsen H, ed. (2000). Toetsen en begrenzen: een ethische en politieke beoordeling van de moderne biotechnologie. ISBN 978-9072016324
- Application-of-amplified-fragment-length-polymorphism-AFLPs-analysis-to-plant-breeding-and-genetics-Procedures-applications-and-prospects-260189595: researchgate.net, Retrieved 11 March, 2020
- Hammond, Scott M.; Caudy, Amy A.; Hannon, Gregory J. (Feb 2001). "Post-transcriptional gene silencing by double-stranded RNA". Nature Reviews Genetics. 2 (2): 110–119. doi:10.1038/35052556. PMID 11253050
- Mutagenesis-in-plant-breeding-for-disease-and-pest-resistance, mutagenesis: intechopen.com, Retrieved 21 August, 2020
- Meudt HM, Clarke AC (March 2007). "Almost forgotten or latest practice? AFLP applications, analyses and advances". Trends Plant Sci. 12 (3): 106–17. doi:10.1016/j.tplants.2007.02.001. PMID 17303467

3 Breeding for Pollinated Species

- **Mass Selection Method for Pollinated Species**
- **Pure-Line Selection Method for Pollinated Species**
- **Hybridisation for Pollinated Species**
- **Breeding for Cross-Pollinated Crops**
- **Breeding for Self-Pollinated Crops**

A range of breeding methods are used in relation to self-pollinated and cross-pollinated species such as mass selection, pure line selection, hybridization, etc. This chapter closely examines the subject of breeding for self and cross-pollinated species to provide an extensive understanding of the subject.

Self-pollinated species have a genetic structure that has implication in the choice of methods for their improvement. They are naturally inbred and hence inbreeding to fix genes is one of the goals of a breeding program for self-pollinated species.

At the beginning of each project, the breeder should decide on the type of cultivar to breed for release to producers. The breeding method used depends on the type of cultivar to be produced. There are six basic types of cultivars that plant breeders develop. These cultivars derive from four basic populations used in plant breeding inbred pure lines, open-pollinated populations, hybrids, and clones. Plant breeders use a variety of methods and techniques to develop these cultivars.

Pure-line Cultivars

Pure-line cultivars are developed for species that are highly self-pollinated. These cultivars are homogeneous and homozygous in genetic structure, a condition attained through a series of self-pollinations. These cultivars are often used as parents in the production of other kinds of cultivars. Pure-line cultivars have a narrow genetic base. They are desired in regions where uniformity of a product has a high premium.

Open-pollinated Cultivars

Contrary to pure lines, open-pollinated cultivars are developed for species that are naturally cross-pollinated. The cultivars are genetically heterogeneous and heterozygous. Two basic types of open-pollinated cultivars are developed. One type is developed by improving the general population by recurrent (or repeated) selection or bulking and increasing material from selected superior inbred lines. The other type, called a synthetic cultivar, is derived from planned matings involving selected genotypes. Open-pollinated cultivars have a broad genetic base.

Hybrid Cultivars

Hybrid cultivars are produced by crossing inbred lines that have been evaluated for their ability to produce hybrids with superior vigor over and above those of the parents used in the cross. Hybrid production exploits the phenomenon of hybrid vigor (or heterosis) to produce superior yields. Heterosis is usually less important in crosses involving self-pollinated species than in those involving cross-pollinated species. Hybrid cultivars are homogeneous but highly heterozygous. Pollination is highly controlled and restricted in hybrid breeding to only the designated pollen source. In the past, physical human intervention was required to enforce this strict pollination requirement, making hybrid seed expensive. However, with time, various techniques have been developed to capitalize on natural reproductive control systems (e.g., male sterility) to facilitate hybrid production. Hybrid production is more widespread in crosspollinated species (e.g., corn, sorghum), because the natural reproductive mechanisms (e.g., cross-fertilization, cytoplasmic male sterility) are more readily economically exploitable than in self-pollinated species.

Clonal Cultivars

Seeds are used to produce most commercial crop plants. However, a significant number of species are propagated by using plant parts other than seed (vegetative parts such as stems and roots). By using vegetative parts, the cultivar produced consists of plants with identical genotypes and is homogeneous. However, the cultivar is genetically highly heterozygous. Some plant species sexually reproduce but are propagated clonally (vegetatively) by choice. Such species are improved through hybridization, so that when hybrid vigor exists it can be fixed (i.e., the vigor is retained from one generation to another), and then the improved cultivar propagated asexually. In seed-propagated hybrids, hybrid vigor is highest in the F1, but is reduced by 50% in each subsequent generation. In other words, whereas clonally propagated hybrid cultivars may be harvested and used for planting the next season's crop without adverse effects, producers of sexually reproducing species using hybrid seed must obtain a new supply of seed, as previously indicated.

Apomictic Cultivars

Apomixis is the phenomenon of the production of seed without the benefit of the union of sperm and egg cells (i.e., without fertilization). The seed harvested is hence genetically identical to the mother plant (in much the same way as clonal cultivars). Hence, apomictic cultivars have the same benefits of clonally propagated ones, as previously discussed. In addition, they have the convenience of vegetative propagation through seed (versus propagation through cuttings or vegetative plant parts). Apomixis is common in perennial forage grasses.

Multilines

Multilines are developed for self-pollinating species. These cultivars consist of a mixture of specially developed genotypes called isolines (or near isogenic lines) because they differ only in a single gene (or a defined set of genes). Isolines are developed primarily for disease control, even though these cultivars could, potentially, be developed to address other environmental stresses. Isolines are developed by using the techniques of backcrossing in which the F1 is repeatedly crossed to one of the parents (recurrent parent) that lacked the gene of interest (e.g., disease resistance).

Genetic Structure of Cultivars and its Implications

The products of plant breeding that are released to farmers for use in production vary in genetic structure and consequently the phenotypic uniformity of the product. Furthermore, the nature of the product has implications in how it is maintained by the producers, regarding the next season's planting.

Homozygous and Homogeneous Cultivars

A cultivar may be genetically homozygous and hence produce a homogeneous phenotype or product. Self-pollinated species are naturally inbred and tend to be homozygous. Breeding strategies in these species are geared toward producing cultivars that are homozygous. The products of economic importance are uniform. Furthermore, the farmer may save seed from the current season's crop (where legal and applicable) for planting the next season's crop, without loss of cultivar performance, regarding yield and product quality. This attribute is especially desirable to producers in many developing countries where the general tradition is to save seed from the current season for planting the next season. However, in developed economies with well-established commercial seed production systems, intellectual property rights prohibit the reuse of commercial seed for planting the next season's crop, thus requiring seasonal purchase of seed by the farmer from seed companies.

Heterozygous and Homogeneous Cultivars

The method of breeding of certain crops leaves the cultivar genetically heterozygous yet phenotypically homogeneous. One such method is hybrid cultivar production, a method widely used for the production of especially outcrossing species such as corn. The heterozygous genetic structure stems from the fact that a hybrid cultivar is the F_1 product of a cross of highly inbred (repeatedly selfed, homozygous) parents. Crossing such pure lines produces highly heterozygous F_1 plants. Because the F_1 is the final product released as a cultivar, all plants are uniformly heterozygous and hence homogeneous in appearance. However, the seed harvested from the F_1 cultivar is F_2 seed, consequently producing maximum heterozygosity and heterogeneity upon planting. The implication for the farmer is that the current season's seed cannot be saved for planting the next season's crop for obvious reasons. The farmer who grows hybrid cultivars must purchase fresh seed from the seed company for planting each season. Whereas this works well in developed economies, hybrids generally do not fit well into the farming systems of developing countries where farmers save seed from the current season for planting the next season's crop. Nonetheless, the use of hybrid seed is gradually infiltrating crop production in developing countries.

Heterozygous and Heterogeneous Cultivars

Other approaches of breeding produce heterozygous and homogeneous (relatively) cultivars, for example, synthetic and composite breeding. These approaches will allow the farmer to save seed for planting. Composite cultivars are suited to production in developing countries, while synthetic cultivars are common in forage production all over the world.

Homozygous and Heterogeneous Cultivars

Examples of such a breeding product are the mixed landrace types that are developed by producers. The component genotypes are homozygous, but there is such a large amount of diverse genotypes included that the overall cultivar is not uniform.

Clonal Cultivar

Clones, by definition, produce offspring that are not only identical to each other but also to the parent. Clones may be very heterozygous but whatever advantage heterozygosity confers is locked in for as long as propagation is clonally conducted. The offspring of a clonal population are homogeneous. Once the genotype has been manipulated and altered in a desirable way, for example through sexual means (since some species are flowering, but are vegetatively propagated and not through seed), the changes are fixed for as long as clones are used for propagation. Flowering species such as cassava and sugarcane may be genetically improved through sex-based methods, and thereafter commercially clonally propagated.

Types of Self-pollinated Cultivars

In terms of genetic structure, there are two types of self-pollinated cultivars:

- Those derived from a single plant.
- Those derived from a mixture of plants.

Single-plant selection may or may not be preceded by a planned cross but often it is the case. Cultivars derived from single plants are homozygous and homogeneous. However, cultivars derived from plant mixtures may appear homogeneous but, because the individual plants have different genotypes, and because some outcrossing (albeit small) occurs in most selfing species, heterozygosity would arise later in the population. The methods of breeding self-pollinated species may be divided into two broad groups those preceded by hybridization and those not preceded by hybridization.

Mass Selection Method for Pollinated Species

Mass selection is one of the oldest methods of crop improvement. In this method, individual plants are selected on the basis of phenotype from a mixed population, their seeds are bulked and used to grow the next generation. In other words, it is a method of breeding in which individual plants are selected on the basis of phenotype from a mixed population, their seeds are bulked and used to grow the next generation.

The main features of varieties developed by mass selection in self and cross pollinated species are given below:

Application

Mass selection is applicable to both self and cross pollinated species. However, it is more commonly used in the improvement of cross pollinated crops than in self pollinated species. This method is rarely used in vegetatively propagated crops.

Genetic Constitution

In self pollinated crops, a mass selected variety is homozygous but heterogeneous, because it is a mixture of several pure-lines. In cross pollinated crops, such varieties are mixture of homo and heterozygotes and are heterogeneous, because they consist of several homo and heterozygous genotypes.

Adaptation

Mass selected varieties have wide adaptation and are more stable against environmental changes due to heterogeneity which provides better buffering capacity. In other words, mass selected varieties have broader genetic base than pure lines. They exhibit more or less stable performance. However, adaptability is more in cross pollinated crops than in self pollinated species.

Variation

They are composed of several pure lines in self pollinated crops and of several homo and heterozygous genotypes in cross pollinated crops. Hence there is heritable variation in the mass selected varieties, besides environmental variation. The heritable variation provides them good buffering capacity.

Selection

Selection is effective in case of mass selected varieties of self pollinated crops due to presence of heritable varieties. However, further selection in the mass selected varieties of cross pollinated crops may lead to inbreeding depression.

Quality

A variety developed by mass selection is less uniform in the quality of seed than purelines due to presence of heritable variation.

Resistance

Mass selected varieties are less prone to the attack of new diseases due to genetic diversity. In other words, they are more resistant or tolerant to new diseases.

Roguing

Periodic removal of off type plants is essential to maintain the yield of mass selected varieties.

Types of Mass Selection

There are two types of mass selection:

- Positive mass selection,
- Negative mass selection.

Positive Mass Selection

When desirable plants are selected from a mixed population and their seeds are mixed together to grow further generation, it is referred to as positive mass selection. This process is continued for several years. Generally, old varieties or land races are used as the base material for mass selection. Selection of desirable plants or positive approach is in common use in mass selection.

Negative Mass Selection

When only undesirable off type of plants are removed from the field and rest are allowed to grow further, it is known as negative mass selection. This is generally used for varietal purification in seed production and certification programs. This helps in maintaining high level of genetic purity in the varieties especially in the self pollinated species.

The success of mass selection mainly depends on three factors:

- Variability in the base population,
- Mode of inheritance of character to be improved,
- Heritability of the character.

Mass selection is more successful in old heterogenous variety or land races than in improved varieties.

Modifications of Mass Selection

Mass selection is practiced both in self and cross pollinated crop plants. However, it is more common in cross pollinated species than in self pollinated species.

There are two defects of mass selection as given below:

No Control on Pollination

In mass selection there is no control on the pollination. The selected plants are pollinated both by superior and inferior pollen parents.

Selection is based on Phenotype

The phenotypic performance is greatly influenced by environmental factors such as soil heterogeneity.

In order to overcome these defects, three modifications of mass selection have been suggested in cross pollinated species.

These modifications are:

- Rejection of inferior pollen plants,
- Use of composite pollen,
- Stratification of the field.

Rejection of Inferior Plants

In this method inferior pollen parents are removed before flowering and rest are allowed for inter-mating. This controls pollination by inferior plants.

Use of Composite Pollen

In this method, pollen is collected from the selected plants and bulked. This composite pollen is used to pollinate the selected plants. This also controls pollination by inferior plants.

Stratification of Field

This is known as grid method of mass selection and was suggested by Gardner. In this method, the field in which mass selection is to be practiced is divided into small plots in such a way that each plot should have 40-50 plants. Now superior plants are selected in each small plot. This eliminates the effect of environment [soil heterogeneity] because small plots are more homogeneous than larger plots.

The success of mass selection is more with oligogenic recessive characters than with polygenie characters. In case of polygenic traits, large number of individual plants should be selected in each generation to tap maximum desirable alleles. Moreover, mass selection is more effective for characters having high heritability. Generally, oligogenic characters have higher heritability than polygenic characters.

In cross pollinated species, the progeny of each cycle of mass selection should be grown in isolation to prevent contamination with other varieties (mating types) of a crop. The isolation distance differs from crop to crop depending upon the agency of pollination.

Generally, more isolation distance is required for wind pollinated species than insect pollinated ones. In general, isolation distance of 300 metres in case of insect pollinated species and 500-1000 metres for wind pollinated species is sufficient. In self pollinated species distance of 5-10 metres is adequate.

Procedure of Mass Selection

Mass selection consists of various steps, viz., selection of base population, selection of desirable plants from base population and mixing their seeds to rise next generation, evaluation in field trials, and releasing as a new variety.

The general procedure of mass selection is given below:

First Year

An unimproved old variety or land race is used as a base population which is grown in a large plot. Then individual plants (55-1000) are selected on the basis of phenotypic performance for characters like height, maturity, disease resistance, productivity etc. The selected plants are harvested at maturity and their seeds are mixed together to grow next generation. This process is repeated till desirable results are achieved.

Second Year

The crop is grown from the bulk seed of selected plants in a separate field using standard variety as a check for comparison of performance. In other words, the material is evaluated in preliminary yield trial. If mass selection is used for purification of old mixed variety, the same old variety can be used as check for comparison.

Third to Sixth Year

The performance of bulk is evaluated for yield and adaptation in main yield trials for 3 to 4 years using standard check for comparison.

Seventh and Eighth Year

The variety is released and named in seventh year and seed is multiplied. In the eighth year the seed is ready for distribution.

Merits and Demerits of Mass Selection

Merits

- This is a good method for improvement of old varieties and land races. This is also used for the purification of improved varieties.
- Mass selected varieties are more stable in their performance than pure-lines. In other words, they have more buffering capacity than pure-lines due to heterogeneity.
- Mass selected varieties provide good protection against diseases.
- Mass selection is a simple and quick method of crop improvement. It takes about 8 years for the release of a new variety, whereas pure-line selection takes about 10 years in the development of new variety.
- This method is applicable to both self and cross pollinated species.

Mass selection has played more significant role in developing new varieties in cross pollinated species than in self pollinated species.

Demerits

- The selection is based on the phenotypic performance. The superior phenotype is not always an indication of superior genotype. The real breeding value of single plants can be judged from the performance of their progeny. Progeny test is not carried out in mass selection.
- In cross pollinated species, there is no control on the pollination. The selected plants are pollinated by both superior and inferior pollen parents. This results in rapid deterioration of variety developed by mass selection.
- In cross pollinated crops, large numbers of plants have to be selected for bulking, because small sample will lead to inbreeding depression.
- The produce of varieties developed by mass selection is less uniform man pure lines. This is because mass selected varieties are mixture of several pure lines in self pollinated crops and consists of several genotypes in cross pollinated species.
- In self pollinated species, pure line selection is more effective than mass selection, because pure line selection leads to isolation of best line from a mixed or heterogeneous population.

Pure-line Selection Method for Pollinated Species

Selection refers to the process that favours survival and further propagation of some plants having more desirable characters than others. The end product of selection process is also known as selection. Natural selection operates in nature without human interference. Natural selection favours those characteristics that are essential for survival of a species.

Thus survival or adaptation is the main concern in the natural selection. Artificial selection, on the other hand is made by human. It favours those characteristics of plants that are related to yield and quality. Thus productivity of economic product is the main concern in human selection. Human selection refers to identification and retention of those plants from heterogeneous populations that are more useful for human than others.

Selection is an important activity in plant breeding programmes. Various selection procedures, such as pure-line selection, progeny selection, clonal selection, recurrent selection, and disruptive selection are commonly used in crop improvement programmes depending upon the mode of pollination of a crop species, the predominant gene action, and the breeding objective.

Selection is practised both in homozygous as well as segregating populations. The effectiveness of selection largely depends on two main factors:

- The extent of genetic variability present in the base population in which selection has to be practised,
- The heritability of the character.

Selection is more effective when sufficient genetic variation is present in the base population. Similarly, selection is more effective for those characters having high heritability than those having low heritability. Selection leads to depletion of genetic variability in a population.

Concept of Pure-line Selection

The concept of pure-line selection was developed in the middle of 19th century in Sweden. However, the genetic basis of pure lines was explained by Johannsen, a Danish biologist in 1903. Pure-line refers to the homogenous progeny of a self pollinated homozygous plant.

Development of new variety through identification and isolation of single best plant progeny is known as pure-line selection or individual plant selection. This method is commonly used in self-pollinated species. Vilmorin in Sweden used this method for the improvement of wheat and barley. The best pure line may be used as a variety or as a parent in the hybridization for the development of superior cultivars.

Features of Pure-line Selection

Pure line selection is practised in heterogeneous populations such as introduced materials, landraces and mass selected varieties of self-pollinated species to isolate superior genotypes. The main features of pure-lines are briefly presented below:

Homogeneous

All the genotypes of a pure-line are homogeneous i.e. genetically identical and phenotypically similar. Thus pure-lines are homozygous and homogeneous.

Non-heritable Variation

The variation within a pure-line is entirely due to environmental factors. Thus the variation is non-heritable in the pure-lines.

Highly Uniform

A variety developed by pure-line selection is highly uniform in quality due to absence of genetic variation.

Selection is Ineffective

Selection is ineffective in a pure-line due to lack of heritable variation. Selection is effective when heritable variation is present.

Narrow Adaptation

Generally, pure-line varieties have narrow adaptation and poor adaptability than heterogeneous populations. The poor adaptability is due to narrow genetic base.

More Prone to New Diseases

Pure-line varieties are more prone to the attack of new diseases due to genetic uniformity and narrow genetic base.

Isolation of Pure-lines

Pure-lines can be isolated from heterogeneous population as well as segregating populations through individual plant selection and progeny testing.

Sources of Variation

In a pure-line variety, natural outcrossing, mutations and mechanical mixtures are the important sources of genetic variation. The spontaneous mutations cannot be controlled. Other two factors can be controlled.

Theory of Pure-line Selection

Johannsen, a Danish biologist, developed the concept of pure-line theory working with Princess variety of common bean (Phaseolus vulgaris).

The common bean is a self-pollinated species. Johannsen isolated 19 different lines on the basis of seed weight form the original seed of the princess variety. Each of the 19 isolated lines had characteristic mean weight of seed. The line No. 1 with largest seed had a mean weight of 640 mg and the line No. 19 with smallest seed showed mean seed weight of 350 mg.

There were seeds of various sizes within each line but the variability within the line was much lesser than the original seed lot. When the seeds in each line were separated into various classes of 100 mg difference and grown separately, the seeds of different classes within a pure-line produced progeny with the same mean weight.

In other words, the largest and smallest seed from a line produced progeny with the same mean weight.

He concluded that (i) continuous inbreeding (selfing) leads to homozygosity, (ii) variation within a pure-line results from environmental factors only, (iii) selection within a pure line is not effective because all the plants in a pure-line have exactly the same genotype, and (iv) selection in the original population is effective because the plants have genetic variation.

The main effect of self-pollination is increase in homozygosity. The homozygosity refers to the proportion of homozygous individuals in a population. Thus there is difference between homozygous and homozygosity. The term homozygous is used for individuals having like alleles on the corresponding locus and homozygosity refers to the proportion of homozygous plants in a population.

The homozygosity is calculated with the help of following formula:

Homozygosity = $[(2^m - 1)/2^m]^n$

where,

- m = Number of generations of self-pollination, and
- n = Number of gene pairs segregating.

This formula is based on two assumptions:

- Equal survival of all genotypes,
- Absence of linkage.

Linkage increases the proportion of homozygous individuals by producing more parental types than recombinants. The final result of self-pollination or selfing is a homozygous but not a homogeneous population, because end product of self-pollination consists of several homozygous lines. The number of various lines that can be obtained by selfing is equal to 2^n, where n is the number of heterozygous gene pairs.

Procedure of Pure-line Selection

Pure-line selection consists of four major steps:

- Selection of a heterogeneous population from which pure-lines have to be isolated,
- Isolation of pure-lines by individual plant selection,
- Testing of pure-lines in field trials,
- Release of the best pure-line as a variety.

The year wise procedure is given as follows:

First Year: An old variety or land race is used for pure-line selection. Single plants are selected from the heterogeneous population keeping in view the objective of selection. The number of individual plants to be selected may vary from 200 to 1000 in various crops.

Second Year: The progeny of each selected plant is grown separately in few rows and evaluated for the character under consideration. The top 15-20 progenies are selected and seed of all plants in each progeny is bulked which constitutes strains.

Third Year: The strains constituted in second years are evaluated in replicated field trials and top performing few strains are selected for further evaluation.

Fourth to Seventh Year: The selected strains are evaluated in field trials for 2-3 years for yield performance. The best genotype is identified on the basis of yield performance.

Eighth to Tenth Year: The best performing strain is released and notified as a variety. Then the breeder, foundation and certified seeds are produced. The production of certified seed takes two years after release of a variety. Thus, the seed of new variety reaches the farmers in tenth year.

Merits and Demerits of Pure-line Selection

The important merits and demerits of pure-line selection are briefly presented below:

Merits

- This is a good method of isolating the best genotype for yield, disease resistance, insect resistance, earliness, quality, etc. from a heterogeneous, or mixed population of an old variety.
- The variety developed by this method is uniform and more attractive than mass selected variety.
- This is an easy and cheap method of crop Improvement.

Superior varieties have been isolated by pure-line selection from the heterogeneous populations in several self-pollinated crops like wheat, barley, paddy, peanut, chickpea, black-gram, green-gram, linseed, cotton and tobacco. Pure-lines are used either as varieties or as parents in hybridization for development of superior varieties or hybrids in self-pollinated species.

Demerits

- This method can isolate only superior genotypes from the mixed population. It cannot develop new genotypes.

- This method is applicable to self-pollinated species only. It cannot be used for development of variety in cross pollinated species.
- The varieties developed by pure-line selection have poor adaptability due to narrow genetic base. All the plants of a pure-line have identical genotypes. Hence, such varieties are more prone to the attack of new disease due to genetic uniformity.

Hybridisation for Pollinated Species

Hybridization is the process of crossing two genetically different individuals to result in a third individual with a different, often preferred, set of traits. Plants of the same species cross easily and produce fertile progeny. Wide crosses are difficult to make and generally produce sterile progeny because of chromosome-pairing difficulties during meiosis.

Hybridization is the process of crossing two genetically different individuals to create new genotypes. For example, a cross between a parent 1, with the genetic makeup (genotype) BB, and parent 2, with bb, produces progeny with the genetic makeup Bb, which is a hybrid (the first filial generation or F_1). Hybridization was the basis of Gregor Mendel's historic experiments with garden peas. Inheritance studies require crossing plants with contrasting or complementary traits.

Hybridization of plants occurs in nature through various mechanisms. Some plants (such as the oil palm) are insect-pollinated, and others (such as maize, or corn) are wind-pollinated.

Natural hybridization has played a significant role in producing new genetic combinations and is the norm in cross-pollinated plants. It is a common way of generating genetic variability.

In plants with perfect flowers (autogamous, having flowers with both stamens and pistils), cross pollination rarely occurs. Such plants (such as wheat and rice) are called self-pollinated plants.

Flowers bearing only pistils or stamens are said to be imperfect flowers. Plants that have separate pistillate and staminate flowers on the same plant (such as maize) are called monoecious. Plants that have male and female flowers on separate plants (such as asparagus) are called dioecious.

Through artificial means (controlled pollination), hybridization of both cross-pollinated and self-pollinated plants can be accomplished. Artificial hybridization is an important aspect of improving both cross-pollinated and self-pollinated plants.

The breeder must know the time of development of reproductive structures of the species, treatments to promote and synchronize flowering, and pollinating techniques.

Objectives of Hybridization

- To artificially create a variable population for the selection of types with desired combination of characters.
- To combine the desired characters into a single individual.
- To exploit and utilize the hybrid varieties.

Types of Hybridization

Hybridization may be of following types:

Intra-varietal Hybridization

The crosses are made between the plants of the same variety.

Inter-varietal or Intraspecific Hybridization

The crosses are made between the plants belonging to two different varieties.

Interspecific Hybridization or Intragenric Hybridization

The crosses are made between two different species of the same genus.

Introgressive Hybridization

Transfer of some genes from one species into the genome of the other species is known as introgressive hybridization. The crosses between different species of the same genus or different genera of the same family are also known as distant hybridization or wide crossing. Such crosses are called distant crosses.

Procedure of Hybridization

It involves the following steps:

- Selection of parents.
- Selfing of parents or artificial self-pollination.
- Emasculation.
- Bagging.

- Tagging.
- Crossing.
- Harvesting and storing the F_1 seeds.
- Raising the F_1 generation.

Selection of Parents

The selection of parents depends upon the aims and objectives of breeding. Parental plants must be selected from the local areas and are supposed to be the best suited to the existing conditions.

Selfing of Parents or Artificial Self-pollination

It is essential for inducing homozygosity for eliminating the undesirable characters and obtaining inbreeds.

Emasculation

It is the third step in hybridization. Inbreeds are grown under normal conditions and are emasculated. Emasculation is the removal of stamens from female parent before they burst and shed their pollens.

It can bc defined as the removal of stamens or anthers or the killing of the pollen grains of a flower without affecting in any way the female reproductive organs. Emasculation is not required in unisexual plants but it is essential in bisexual or self-pollinated plants. Various methods used for emasculation are:

Hand Emasculation or Forceps or Scissor Method

This method is generally used in those plants which have large flowers. In this method the corolla of the selected flowers is opened and the anthers carefully removed with the help of fine-tip forceps.

Following are the important precautions while performing this method:

- Flowers should be selected at proper stage.
- Stigma should be receptive and anthers should not have dehisced.
- All the anthers should be removed from the flowers without breaking.
- Stigma and ovary of the flower should not be damaged.

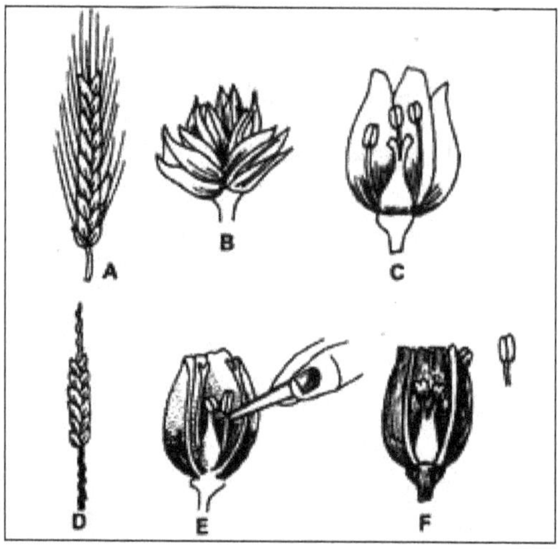

(A-F) Emasculation in wheat (A) Spike of spikelets (B) spikelet, (C) Floret,(D) Upper and lower spikelets removed, awans removed, upper portion of florets cut, (E) Anthers removed with the help of fine-tip forceps. (F) Removed anther.

Hot Water Treatment

Removal of stamens with the help of forceps is very difficult in minute flowers. In such small hermaphrodite flowers (e.g., Bajra, Jowar) emasculation is done by dipping the flowers in hot water for a certain duration (1-10 minutes) of time.

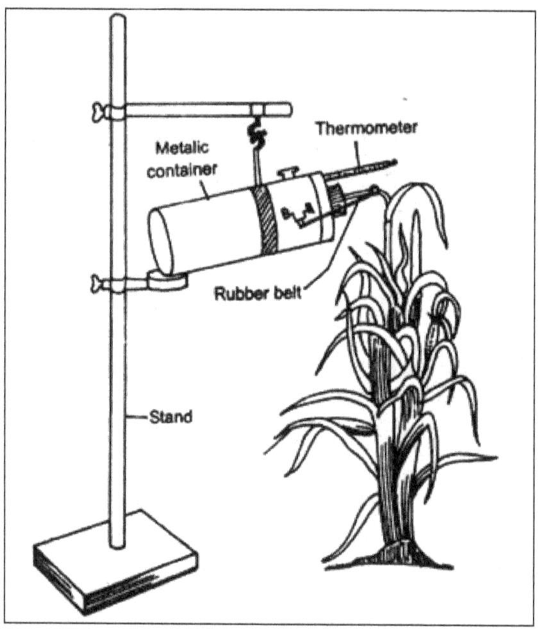

Hot water equipment emasculation.

The time varies from species to species. This method Is based on the fact that gynoecia can withstand the hot temperature at which the anthers are killed. In this method an equipment is used which is placed on a simple heavy stand.

It consists of a cylindrical metallic container of 60 cm length, with one hole of 5 cm to 16 cm diameter on one end to pass over a bajra or jowar head. After inserting the panicle inside the container a cork is fitted in the hole to close it.

A 35 cm long rubber tube or belt is stretched over the side of the container, and when in use this tube is tied around the peduncle of the head. To measure the temperature, in the upper side of the container a thermometer is placed. In the field water is carried in a thermos jug.

The panicle is inserted in the container prior to blooming for a particular duration of time. It has been observed that pollen grains of rice are killed by immersing the inflorescence for 5 to 10 minutes in the hot water maintained at 40-44°C in a thermos flask.

Cold Water Treatment

Like hot water cold water also kills pollen grains without damaging the gynoecium. In rice 0-6°C temperature is maintained to kill the pollen grains. This method is less effective than hot water treatment.

Alcohol Treatment Method

This method is not commonly used for emasculation because duration of treatment is an important factor since a very short duration is required failing which even the gynoecium may be damaged. Flowers or inflorescences are immersed in alcohol of a suitable concentration for a brief period. In alfa-alfa, a treatment of even 10 seconds with 57 % alcohol is sufficient to kill the pollen grains.

Suction Method

It is a mechanical method and is suitable for the crops having minute flowers. In this method the amount of pressure is applied in such a way that only anthers are sucked out and other parts of the flower like gynoecium remain intact. However, in this method 10-15% self pollination takes place. It is one of the major drawback of this method.

Male Sterility or Self-incompatibility Method

Emasculation option can be eliminated by the use of male-sterile plants, In some self-pollinated plants for example, Sorghum, Onion, Barley etc. anthers are sterile and do not produce any viable pollens. Similarly self-incompatibility may also be used to avoid emasculation.

Chemical Gametocides

Certain chemicals are capable of causing male sterility, when sprayed before flowering e.g., 2, 4-D, naphthalene acetic acid (NAA), maleic-hydrazide (MA), tribenzoic acid etc. FW450 in cotton may be used for bringing about emasculation.

Bagging

It is the fourth step and completed with emasculation. The emasculated flower or inflorescence is immediately bagged to avoid pollination by any foreign pollen. The bags may be made of paper, butter paper, glassine or fine cloth. Butter paper or vegetable parchment bags are most commonly used.

The bags are tied to the base of the inflorescence or to the stalk of the flower with the help of thread, wire or pins. The bagging is done with the emasculation in bisexual plants and before the stigma receptivity and dehiscence of the anthers in unisexual plants. Both male and female flowers are bagged separately to prevent contamination in male flowers and cross-pollination in female flowers.

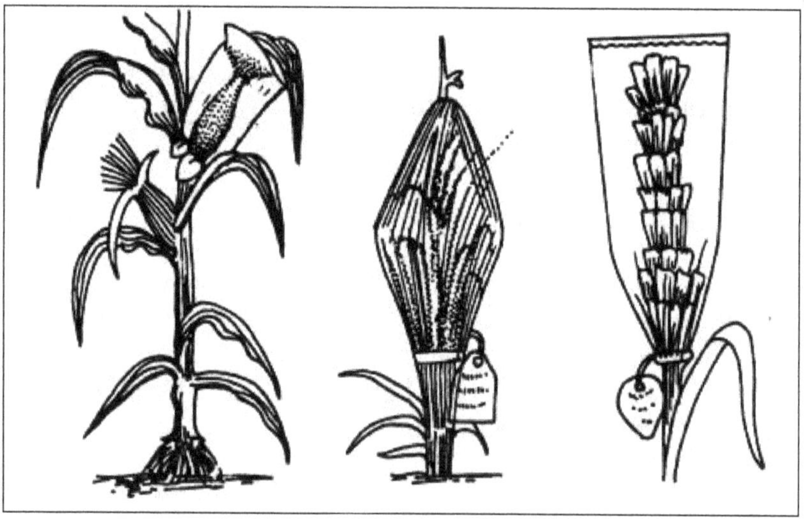

Different methods of Bagging.

Tagging

The emasculated flowers are tagged just after bagging. Generally circular tags of about 3 cm or rectangular tags of about 3 x 2 cm are used. The tags are attached to the base of flower or inflorescence with the help of thread.

The information on tag must be as brief as possible but complete bearing the following information:

- Number referring to the field record.
- Date of emasculation.
- Date of crossing.
- Name of the female parent is written first followed by a cross sign (x) and then the male parent, e.g., C x D denotes that C is the female parent and D is the male parent.

Crossing

It is the sixth step. It can be defined as the artificial cross-pollination between the genetically unlike plants. In this method mature, fertile and viable pollens from the male parent are placed on the receptive stigma of emasculated flowers to bring about fertilization.

Pollen grains are collected in petridishs (e.g., Wheat, cotton etc.) or in paper bags {e.g., maize) and applied to the receptive stigmas with the help of a camel hair brush, piece of paper, tooth pick or forceps. In some crops (e.g., Jowar, Bajra) the inflorescences of both the parents are enclosed in the same bag.

Harvesting and Storing the F_1 Seeds

Crossed heads or pods of desirable plants are harvested and after complete drying they are threshed. Seeds are stored properly with original tags.

Raising the F_1 Generation

In the coming season, the stored seeds are sown separately to raise the F_1 generation. The plants of F_1 generation are progenies of cross seeds and therefore are hybrids.

Hybridization Methods of Plant Breeding in Self-pollinated Groups

There are several methods of improvement of self-fertilized crops by hybridization. These are:

- Single seed descent method.
- Back cross method.
- Multiple cross method.

Mass Pedigree Method

It is a modified form of pedigree method in which segregating material is handled by bulk (mass) method when conditions are unfavorable for selection and by pedigree method when conditions are favourable for selection.

Single Seed Descent Method

This method was suggested by Coulden for advancing segregating generation of self-pollinated crops. A breeding procedure used with segregating populations of self-pollinated species in which plants are advanced by single seeds from one generation to the next is referred to as single seed descent method.

The procedure is as follows:

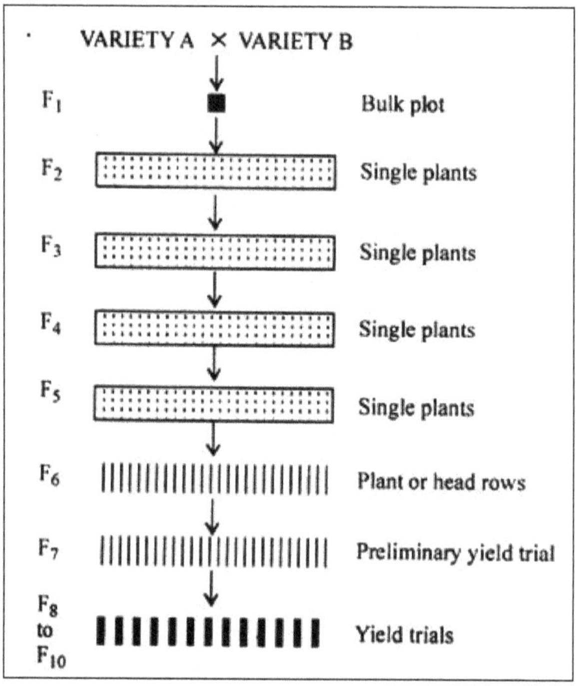

Single-seed-descent method of selection.

Back Cross Method

This method was first proposed by Harlan and Pope. Now-a-days this method is employed in improvement of both self and cross-pollinated crops where varieties are deficient in one or two aspects. This method is used particularly for transferring a single simply inherited character like disease, frost or drought resistance and earliness from an undesirable variety to a good commercial variety.

The desirable variety is called as recurrent or recipient parent and it is crossed to an undesirable variety, called as donor or non-recurring parent (called donor because the desirable genes are transferred).

F_1 plants instead of permitting to self-pollination as in pedigree or bulk method care crossed with the recurring parent and therefore, it is called as back cross method (A back cross may be defined as a crossing of F_1 hybrid with any of its parents).

The procedure is as follows:

Suppose there are two varieties A and B. A is very good in all characters but disease susceptible and the variety B is disease resistant but very poor in all characters. It is needed to transfer the disease resistance from B to A without adversely affecting the good qualities of A. B is then donor or non-recurring parent and A is recipient or recurring parent.

Selected plants of A and B are crossed to raise the F_1 Here A is treated as the female and B as male. F_1 plant is backcrossed with the recurrent parent; variety A and BC_1 plants (back cross first generation) plants are raised. Here B is treated as female and A as male. Select the BC_1plants possessing desired characters of A with disease resistance of B.

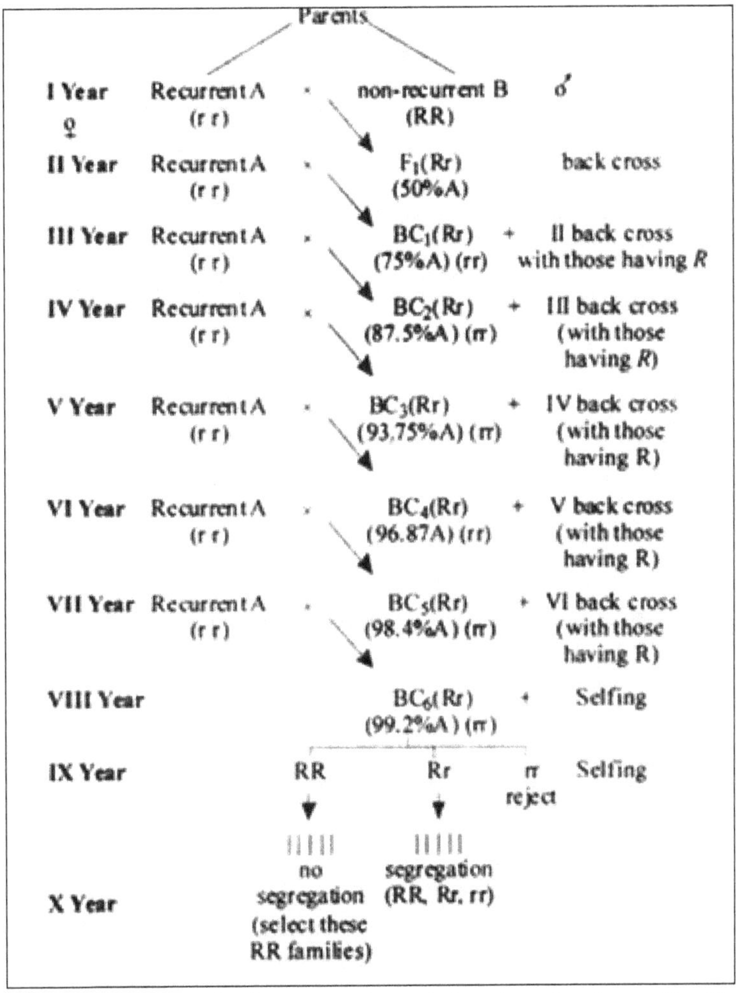

Different steps involved in the back cross method of plant breeding in a self pollinated crop, when a dominant trait is transferred from variety B to an otherwise improved variety A.

The selected BC_1 plants again backcrossed to A till a desirable type having good qualities of A and disease resistance of B is obtained. Disease resistant plants are selfed and individual plant progenies from the selfed seeds of selected plants are grown.

New variety is tested in replicated field trials along with the variety A as check. Seeds are multiplied and released for cultivation. The method is used for the transfer of a dominant gene. In case of recessive gene transfer, a slightly different procedure is followed. It is summarized in figure.

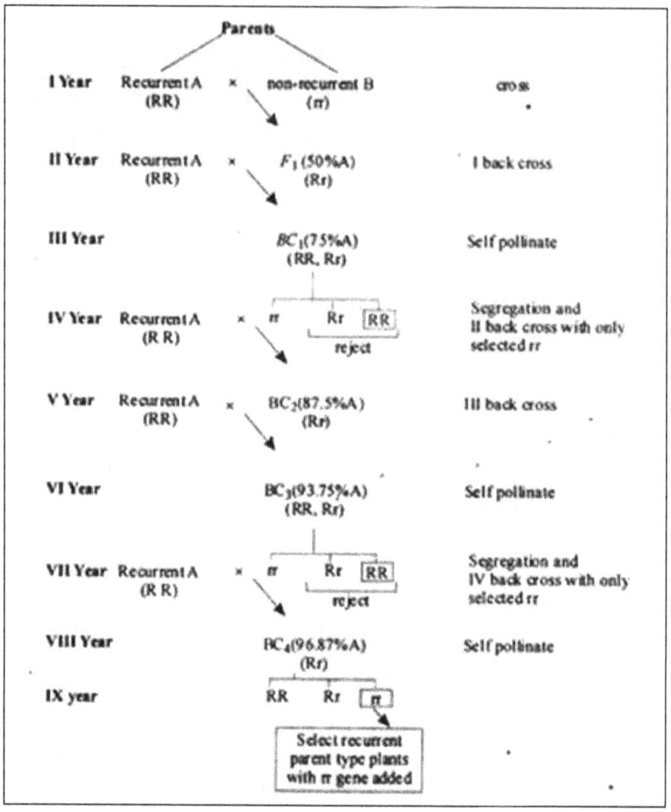

Different steps involved in the back cross method of plant breeding in a self pollinated crop, when a recessive trait is transferred from a variety B to otherwise iproved variety A.

Merits

- This is the most common method used for transfer for oligogenic character like disease resistance.
- It is the only method used for the interspecific transfer of characters, transfer of qualitative characters, quantitative characters, transfer of cytoplasm, particularly for male sterility and for the production of isogenics lines.
- Role of environment is almost negligible. So time required for developing new variety may be drastically reduced because off-season nurseries and greenhouses can be used for developing 2-3 generations each year.
- Extensive yield tests are not required because the performance of the recurrent parent is already known.
- Breeder does not require to handle large populations.
- In this method all desirable characters of a popular variety (genotype known) are retained except the undesirable genes at a particular locus is replaced. So,

the outcome of this method is known forehand, and plants can be reproduced any time in future.

Demerits

- The new variety is superior over the already available variety only in the transferred character.
- Hybridization has to be done for each backcross. It requires lot of time and money.
- Sometimes undesirable genes are closely linked with desirable one and get transferred to the new variety.

Multiple Cross Method

A cross-involving more than one inbred line is referred to as multiple cross. It is also known composite cross and is used to combine monogenetic characters from different sources into a single genotype. In this method, several pure lines are crossed together. The selected pure lines are first combined into crosses as A × B, C × D, E × F, G × H and so on.

The F_1 plants are mated together as (A × B) × (C × D) and (E × F) x (G × H). Finally, the F_1 plants of double crosses are crossed with each other to produce hybrids [(A × B) x (C x D)] x [(E x F) x (G x H)]. Further breeding in these hybrids is carried out according to either pedigree or bulk method.

A × B   C × D   E × F   G × H
AB × CD   EF × GH
ABCD   ×   EFGH
ABCDEFGH

Merits

- In self-pollinated crops this method is used when three or four monogenic characters scattered in three or four different varieties are to be combined into one.
- These crosses generally have wider adaptation.

Demerits

- These crosses are generally less productive.
- This method has limited utility except in high risk areas where severe disease damage occurs regularly from a highly specialized disease pathogen.

Self pollinated crops	
When the pollen of the same flower or other flowers of the same plant pollinates a receptive stigma, the condition so created is called self pollination.	
Annual sweet clover	Lespedeza species
Apricot	Lettuce
Barley	Lima Beans
Bean	Linseed
Brinjal	Mountain Bromegrass
Bur clover	Moong (Green Grams)
Chick peas	Nectarine
Cotton	Oat
Cowpea	Pea
Foxtail millet	Peach
Groundnut or peanut	Pepper (Capsicum annum, C. frutescenes)
Lady's Finger (Okra)	
Potato	Sweet pea
Rice	Tobacco
Soft chess	Tomato
Sorghum (h)	Urd
Soyabean	Velvet bean
Strawberry clover	Vetch
Subterranean clover	Wheat

Hybridization Methods of Plant Breeding in Cross-pollinated Crops

Like self-pollinated crops, hybridization in cross-pollinated crops involves crossing of two or more inbreeds. The inbreeds are combined in any one of the following types of crosses and released as improved strains:

Single Cross

Crossing of two inbreeds or varieties are called single cross such as A x B or C x D. This cross was proposed by Shull. The total number of possible different single crosses produced by a given number of inbreeds can be calculated by the formula n(n-1)/2 where n is the number of inbreeds. e.g., from four inbreeds six single crosses can be produced.

It can be calculated as:

$$\text{No of single crosses} = 4\frac{(4-1)}{2}$$
$$= \frac{4\times 3}{2}$$
$$= 6$$

Single crosses give the maximum degree of hybrid vigour but due to weak inbreeds, little amount of seed is produced.

Double Cross

In this method two single hybrids are crossed together to produce a double cross hybrid for e.g., if A, B, C and D represent inbred lines, one of the possible single cross can be represented by A x B and are of the possible double crosses by (A x B) x (C x D). The be number of double crosses can calculated by the formula $\frac{n(n-1)(n-2)(n-3)}{8}$ where n is the number of inbreds. Thus with 4 inbreds 3 double crosses hybrids can be produced. It can be calculated as follows:

$$= \frac{4(4-1)(4-2)(4-3)}{8}$$
$$= \frac{4\times 3\times 2\times 1}{8}$$
$$= 3 \text{ double crosses hybrids.}$$

These are:

- (A x B) x (C x D).
- (A x C) x (B x D).
- (AxD)x(BxC).

Commercial hybrids are produced by this method. These hybrids give very high yield in small land without any increase in the cost of production. All the hybrid seeds of maize to farmers for cultivation are nothing but double crosses.

Three-way cross:

It is a cross of a single cross (used as female) with another inbred (used as male), i.e., it involves three inbreeds [(A x B) x C], The single cross and the inbreeds are planted in the same way as the inbreeds in the single cross. The advantage of this cross is the use of vigorous hybrid of first generation as female in order to maximize the yield of hybrid seeds as well as to obtain seeds of normal grain size.

Top Cross or Inbred Variety Cross

It can be defined as a cross between an open pollinated variety and inbred line. Either the inbred or the variety may be used as female parent, but to use variety as female is preferable. It is used not only for the developing hybrid but to test the combining ability of the inbreeds.

Synthetic Cross

It can be defined as a cross among inbreeds, clones or sibbed lines without control of pollination. (Synonyms, poly-cross and strain building). This is the cross of a number (4-10) pretested hybrids and is done by open pollination in isolation.

Seeds of these inbreeds are mixed in equal proportion and sown in isolated plots. Natural cross pollination takes place and the harvested produce becomes synthetic cross. This cross is used in forage cross where floral structure causes difficulties in artificial pollination.

After any one of the above-mentioned crosses, seeds are tested, multiplied and distributed. The testing is done at various regional research stations by various research workers. The entire process takes about 5-7 years.

Cross pollination crops	
When the pollen of a flower pollinates the stigma of another flower located on a different plant, it is termed as cross pollination	
Alfalfa or Iuceren	Hemp
Almond	Hops
Alsike Clover	Kale
Apple	Kohlrabi
Artichoke	Maize
Asparagus	Mangel
Banana	Mango
Beet	Muskmelon
Birds foot trefoil	Olive
Blackberry	Onion
Blueberry	Orchard grass
Broccoli	Parsley
Brussels sprout	Palestine
Cabbage	Papaya (papita)
Carrot	Pear
Castor	Pearl millet or bajra
Cauliflower	Pecan
Celery	Plum
Citrus	Pumpkin

Chard	Raspberry
Cherry	Red clover
Chicory	Rhubarb
Chinese cabbage	Rye
Chestnut	Safflower
Collard	Smooth bromegrass
Crimson clover	Spinach
Cucumber	Squash
Date palm	Strawberry
Fig	Sunflower
Filbert	Timothy
Graph	Turnip

Hybridization Methods of Plant Breeding in Vegetatively Propagated Crops

Improved clones of asexually propagating plants are selected and allowed to grow under conditions, which favour flowering and seed setting. Desirable clones are crossed together and the hybrids are multiplied by cloning. Each F_1 plant is a potential source for a new clone. This procedure has been used for developing many improved varieties of sugarcane, potato etc.

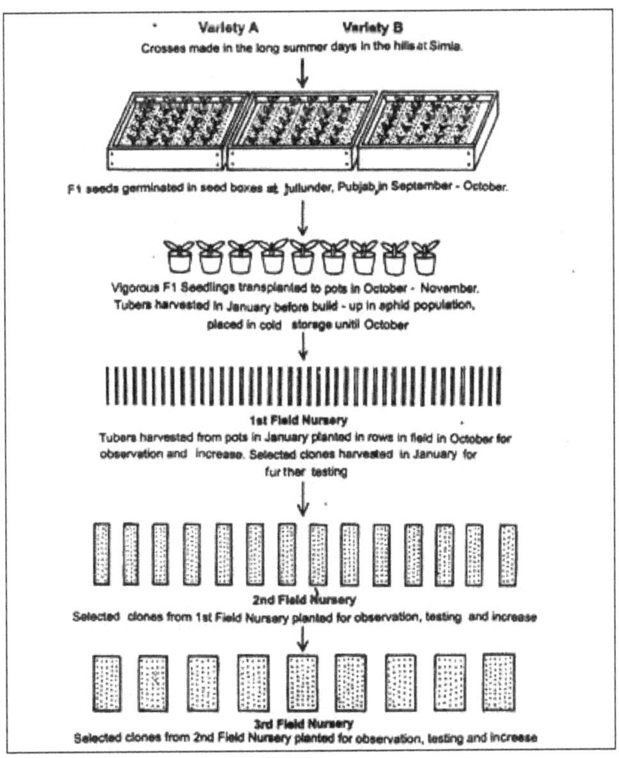

Scheme for growing F_1 potatoplants following hybridization and for screening and testing the clones.

Breeding for Cross-pollinated Crops

Production of Single Cross Hybrids

Crossing between two inbreds or varieties is called single cross. Such crosses express maximum amount of heterosis, but if the inbreds are poor then very little amount of seeds are produced. So, it is not widely used for commercial production.

Total number of single crosses by a given number of inbreds (n) can be calculated by the formula n(n-1) and excluding reciprocals by n(n-1)/2. For example, with 4 inbred lines, 12 single crosses and 6 straight crosses can be made. For commercial production of single crosses, a male sterile line is used as female, the male parent is the restorer line, the seed set in the female parent is the hybrid seed, but the seeds of male parent are selfed seed.

Production of Double Cross Hybrids

Crossing between two single cross hybrids is called double cross, which also involves four different inbred lines. For example, if A, B, C, D represent 4 inbred lines, one of the possible double crosses will be the crossing between two single crosses, A x B and C x D.

The number of possible double cross hybrids can be calculated by the formula.

n(n-1)(n-2)(n-3)/8, where n is the number of inbred lines. So with 4 inbred.

lines, 4x3x2x1/8= 3 double cross hybrids can be produced.

Two Different Schemes can be Taken for Production of Double Cross Hybrids.

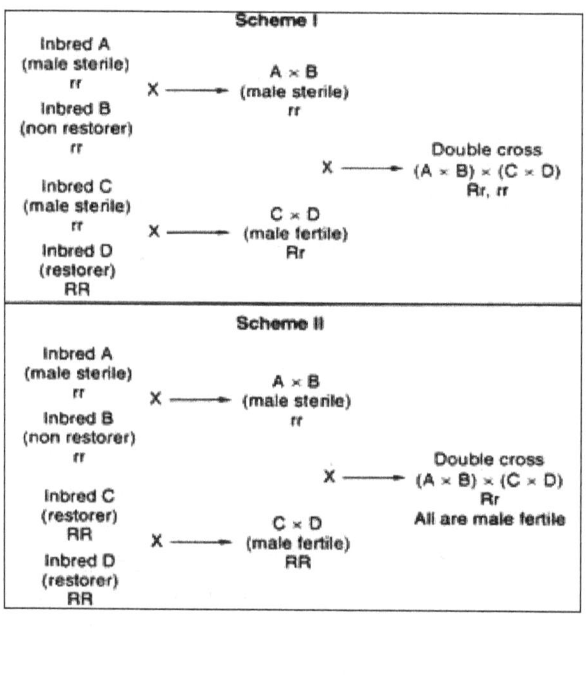

Production of Top-cross Seed

This scheme is also known as inbred-variety cross, (Inbred x Variety), i.e., inbred line is crossed with an open pollinated variety. Also the cross between a single cross and with another inbred is known as double top cross or 3 ways cross.

This method is used not only for developing a hybrid but to test the combining ability of the inbred. The variety or the single cross product is used as male parent and the inbred line to be tested is used as female parent (emasculated), the seeds from the inbred line are harvested.

Production of Synthetic Cross

A synthetic variety is produced by crossing in all combinations a number of lines that combine well with each other, this synthetic variety is maintained by open pollination in isolation.

It is produced by open pollination between several inbred lines together in isolated place. Seeds of these inbreeds are mixed in equal proportion and sown in isolated plots, natural cross pollination takes place and the seeds are harvested together.

The term poly cross is used in case of forage crops where crossing is difficult and also it differs from synthetic cross that the inbred lines are not pretested. Another term composite variety is used as synonym for synthetic variety but by definition it is the production of mixture of several phenotypically similar outstanding lines by open pollination.

Breeding for Self-pollinated Crops

The results of genetic breeding have been excellent for most self-pollinated species under improvement. The evidence is the gain in productivity over the past 30 years. The mean gain of some of the most important species in the country was 4.85 % per year. Assuming that at least 50 % of this progress must be attributed to genetic improvement, the achievement of genetic gain is easily demonstrated. It should be emphasized that the evaluation of the breeding success should not only be based on the yield increase, but, mainly on the improved sustainability of self-pollinated species grown. This increase in sustainability can be measured by the availability of cultivars adapted to new growing conditions, more tolerant to biotic and abiotic stresses.

Among the self-pollinated species there are some as soybean that attract the attention of large multinational seed companies. The case of tomato is also very peculiar, because attention is focused on the production of hybrid seeds and although there are public institutions dedicated to tomato improvement, most of the seed stock is produced by multinational companies. Contrastingly, in the case of some species such as common bean, the development of new cultivars is controlled by the public sector. In essence, the improvement process in different species is rather similar, but there are differences in the resources invested in breeding, especially in terms of the crosses number, progenies evaluated annually and the number of test sites of new lines.

The commonly used breeding methods of self-pollinated plants were developed over a century ago in Europe. No great subsequent changes were made. Basically, the selection procedure of the best progenies was optimized, particularly with the development of agricultural experimentation based on the publications of Fisher. In the past 25 years, the techniques of molecular biology opened up the prospect of the possibility of direct selection of a genotype. The emphasis of global research was focused mainly on the use of molecular DNA markers. Abundant basic information was compiled, but evidently the phenotypic analysis in the field, as accurately as possible, is still indispensable.

Information was accumulated that indicated that the traits the breeders work on are mostly complex, in other words, probably controlled by several genes. This observation became even patent because the improvement consists, in summary, of an accumulation of advantages through successive selection cycles, recurrent selection, that involves intercrosses of the best lines/cultivars available in each cycle. Numerous studies have focused on improving breeding efficiency.

Use of Mixed Models to Improve Selection Efficiency

The mixed model methodology was proposed by the American statistician Charles Roy Henderson in 1949, and has since been routinely used, especially in animal breeding. In plant breeding, this statistical technique only became known in the late twentieth century. A series of advantages are attributed to its use.

Use of Mixed Models in the Pedigree Method

One of the advantages of the pedigree method is the possibility to record the genealogy of all plants/progenies. Although this advantage is often cited, its usefulness is limited. One possibility of exploiting information on the genealogy was pointed out by Nunes For this purpose, a genetic model without epistasis was considered and one trait controlled by 20 segregating genes with independent distribution. Equal gene effects and an allelic frequency of 0.5 were assumed. In the simulation scenarios, different heritability (h^2) values were considered for selection based on the mean of F_4 progenies. For each heritability, 1000 segregating populations with the pedigree method were simulated, with records of the genealogy.

The pedigree method is based on a segregating F_2 population consisting of 64 plants, which resulted in 64 $F_{2:3}$ progenies, each with 40 plants. From each $F_{2:3}$ progeny, two plants were randomly taken, resulting in 128 $F_{3:4}$ progenies. The process was repeated in the following generation in order to obtain 256 $F_{4:5}$ progenies, each with 40 plants.

The phenotypic data from 256 F4:5 progenies of each population were then analyzed in a completely randomized design, with two replications and plots of 20 plants. The mixed model approach was used for this analysis, considering the following models: GA model, which includes information on parentage among the families, and the GI model, in which this information of genetic relationships is ignored. From the above models, the predictions BLUP (Best Linear Unbiased Prediction) with (BLUPA) and without parentage (BLUPI) were determined. It is commonly comment that BLUP-predicted breeding values are free from estimation errors. The simulation showed that the correlation between the predicted additive genetic values (AGV) and the true AGV varied according to the h2 of the trait, i.e., the greater the h^2, the higher the correlation. It is understood, however, that the predictor considering the genealogy (BLUPA) was, independent from h^2, which best predicted the true AGV, i.e., a higher correlation between BLUPA and AGV. The difference between the correlations of predictors BLUPA and BLUPI with AGV were more expressive at lower h2. On the other hand, this result shows that at heritability values of more than 50 %, the genetic values associated with the progenies are already well-determined by phenotypic means, which is equivalent to the BLUPI in terms of selection in balancing condition.

In general, the simulation showed that the inclusion of genetic similarity among progenies in BLUPA resulted in higher gain with selection (GS) than with BLUPI for all pre-established heritability values and proportion of selection. The relative GS increased with the reduction in heritability and selection intensity, i.e., the weight of relatedness was more pronounced at lower heritability. Thus it is evident that the effort of recording the genealogy of the segregating population should be compensated by using this information in the selection of progenies in the application of mixed models, which can result in additional gain from selection.

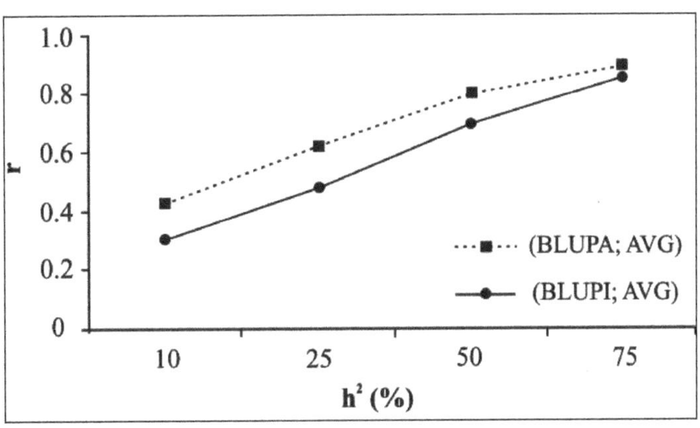

Mean values of the Spearman correlations between the additive genetic values (AGV) and BLUP predictions/estimates considering (BLUPA) and ignoring (BLUPI) AGV parentage between progenies F4:5, carry by the genealogical/pedigree method, for different heritability values in the progeny means (h^2).

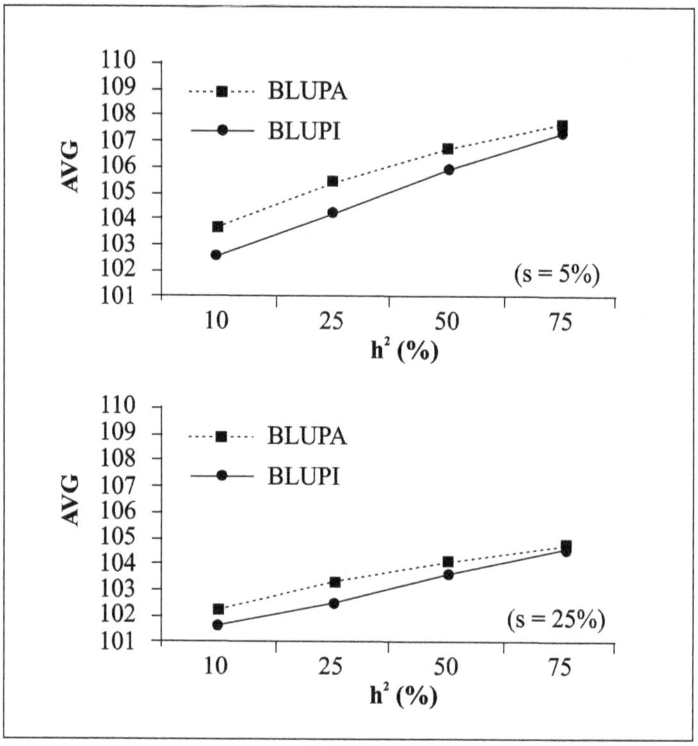

Additive genetic values (AGV) progeny means F4:5 selected by BLUP considering (BLUPA) and ignoring (BLUPI) a information on additive parentage due to the broad-sense heritability in the progeny means (h^2) and selected proportions s of 5 % and 25 %.

Sequential Analysis of the Evaluation Progeny Tests

A recurrent selection program usually begins with a large number of $F_{2:3}$ or $S_{0:1}$ progenies that decreases in every generation. At the end of the process, five to ten plants are selected to participate in a competition experiment of lines for two to six seasons, in various environments, where lines are then identified to participate in tests of VCU (Value for Cultivation and Use). In this situation there are at least two stages of selection, i.e., of the segregating $S_{0:1}$, $S_{0:2}$, $S_{0:3}$, $S_{0:4}$, $S_{0:5}$ progenies. Often only the generation of reference is considered, i.e., in the selection of So:4, for example, the performance of the progenies in the previous generation is not considered.

In the analyses of a decreasing number of progenies, the sequential analysis, which includes all progenies evaluated by BLUP, is suggested as a good alternative. To quantify the improvement in the process efficiency through sequential analysis, was used an evaluation experiment of progenies of two cycles of a recurrent selection program, of the Universidade Federal de Lavras (UFLA) since 1990. The $S_{0:1}$ to $S_{0:10}$ progenies were evaluated in these experiments. The number of progenies ranged from 322 in $S_{0:1}$ to 10 in $S_{0:10}$. Two controls were used in all experiments.

Firstly, the analysis of variance per generation was performed. The subsequent analyses were based on the sequential procedure, recovering the information from the evaluated generations. For this purpose, three analysis strategies were adopted: a) sequential analysis considering a combination of the generations $S_{0:1}$ to $S_{0:3}$, $S_{0:1}$ to $S_{0:4}$ and $S_{0:1}$ to $S_{0:5}$, using all progenies; b) sequential method comprising the 23 best progenies in common in the $S_{0:1}$ to $S_{0:5}$ generations; c) sequential analysis involving the 10 progenies of the $S_{0:1}$-$S_{0:10}$ evaluations. Analyses were performed using Proc Mixed of SAS considering the progeny effect as random (BLUP) and fixed (BLUE). To quantify the efficiency of the analysis procedures the coincidence of progenies selected by the strategies was contrasted with the standard adopted by the breeding program of the institution. The correspondence in the ranking of the progenies in the different analysis strategies was compared as well.

The coincidence between the progenies selected by the traditional method, based only on the mean of the last generation evaluated, was compared with the BLUP considering all generations from $S_{0:1}$, called BLUPT in this study. For the 119 progenies selected in $S_{0:2}$, selected by the mean, coincidence by BLUPT would be 84 in the $S_{0:1}$ and $S_{0:2}$ generations. In the same cycle, now considering the 62 progenies selected in $S_{0:3}$, the coincidence by BLUPT ($S_{0:1}$, $S_{0:2}$ and $S_{0:3}$ generations) would be 39 progenies. The data in Table clearly show that there was little correspondence between the sequential analysis involving all progenies, compared to the traditional procedureThe percentage of this discrepancy in the progeny classification decreases as the number of generations involved increases. In the traditional process, the progenies x generations (seasons) interaction is not considered because only environments or locations within a same generation are considered. Part of this no coincidence should therefore be attributed to the progenies x generations interaction.

Table: Number of selected progenies in the reference generation by the mean and that would coincide considering sequential analysis by BLUPT.

Reference generation	Number of selected progenies in the mean of the reference generation	Number of progenies identified by BLUPT concerning mean
$S_{0:2}$ 1/	119	84(71%) 2/
$S_{0:3}$	62	39(63%)
$S_{0:4}$	23	9(40%)

1 BLUPT analysis considering from $S_{0:1}$ to the reference generation; 2 Percentage of coincidence in relation to the number of selected progenies in the reference generation.

In the $S_{0:5}$ generation, 23 progenies were evaluated that originated the 10 that were evaluated up to $S_{0:10}$. Using some analysis strategies the coincidence of the progeny classification based on the standard procedure was compared with the BLUP and BLUE analyses involving only the 23 progenies, but including information from $S_{0:1}$. It was stated that the coincidence between these two procedures was very high. This shows that the influence of considering a fixed or random progeny effect on the classification was rather low. The same observation was reported in a study of Piepho and Möhring, in which the authors identified a correlation between the results of analysis with BLUP and BLUE in all cases ranging from r = 0.994 to r = 1.0.

A breeding program should be focused mostly on the identification of the best line s in the final evaluations.

Ranking of progenies evaluated considering the reference generation (mean), analysis com information recovery for random effect involving the progenies in common of the generations (BLUP), analysis with information recovery for random effect involving all progenies in all generations (BLUPT) and analysis with information recovery for fixed effect (BLUE) for the 10 progenies evaluated in the elite assays.

Progenies	Ranking			
	Mean	Blue	Blup	Blupt
M-85-17	1	5	5	9
M-55-14	2	3	3	4
M-55-3	3	1	1	1
M-39-18	4	6	6	6
M-85-1	5	4	4	-
M-215-10	6	2	2	3
M-119-4	7	9	9	-
M-85-11	8	7	7	-
M-39-24	9	8	8	-
M-45-5	10	10	10	-

Ten lines were evaluated for five generations (seasons), totaling 13 environments. If one considers the selection of the five best lines for VCU, the correlation was four of five, or 80 %. Interestingly, even when only 10 progenies were considered and in the mean all 13 environments, the coincidence with BLUP and BLUE was greater, but still there was a loss of efficiency.

When considering BLUPT, two of the five best progenies coincided. It should be emphasized, however, that the progenies M-55-14 and M-55-3, classified on average as second and third, would be covered in all procedures.

It was concluded that in the traditional analysis, based on the means only, efficiency is reduced. This loss is more pronounced in the early generations up to $S_{0:5}$, since the progeny x crop interaction is not taken into account. In the later generations, that is $S_{0:6}$ up to $S_{0:8}$ or $S_{0:10}$, when the number of progenies is lower and the analyses are based on all evaluated generations, the efficiency loss is lower. The use of any statistical analysis procedures, with the computer facilities available nowadays, does not entail any additional onus. It is therefore recommended that, in the future, the sequential analysis should be used in recurrent selection programs due to its greater selection accuracy, as suggested by Resende.

Comparison of BLUP with Selection among and within Progenies and Mass Selection to Select Plants of a Segregating Population

There are some alternatives for selection in a segregating population of self-pollinated plants, including the Bulk method and Bulk within F_2 progenies. By these methods, the populations are advanced for a few generations, usually to F_4 or F_5, when plants are selected and their progenies derived are evaluated more detailed in replicated experiments.

There are several options that can be used to identify the plants/progenies in the subsequent selection process, for example, mass selection and selection among and within families, the latter when using progenies from F_2 plants. Another not much explored possibility in common bean breeding is the BLUP procedure. Mendes compared alternatives to identify the best plants in the populations by the method Bulk within F_2 progenies using the population of cross CVIII8511 x RP-26. Of this population, 51 $F_{2:3}$ progenies were obtained and evaluated in the $F_{2:4}$ generation in a randomized block design, with 20 replications and one plant per plot. Yield data per plant and the grade of architecture on a 1 to 9 scale were obtained. The data for each trait were standardized, resulting in a Zindex, and subsequently, the ΣZ, underlying the simultaneous evaluation of the two traits. For this analysis, the software Selegen Computer Program for Genetic Selection, model 59 was used, by which individual genotypic values can be calculated. The 100 best and 100 worst plants were identified, leading to the $F_{4:5}$ progenies. The $F_{2:4}$ generation was also analyzed by the least squares (LS) method, in a randomized block design with 20 replications and one plant per plot. With data from the individual genotypic values, obtained by BLUP, the following strategies were compared for selection in the $F_{2:4}$ generation: a) Mass selection: the 100 best and worst plants were selected in the $F_{2:4}$ generation as a function of ΣZ, regardless of the progeny or replication to which they belonged; b) Stratified mass selection: the plants were divided into strata and each stratum was considered a replication. Thus, each stratum contained one plant of each progeny, with a total of 51 plants per stratum. The five best and worst plants of each replication were selected as a function of ΣZ, resulting in 100 progenies from each group in the 20 replications; c) Selection among and within progenies: from the analysis of variance using the least squares (LS) method, the means of 51 progenies and the six best and worst, respectively, were selected (selection intensity among progenies of 11.7 %) and within these, the 16 best of the 20 existing plants. The same was done for the group of the worst, totaling 96 plants of each group.

The individual BLUP analysis provides an estimate that involves all model variables, for example, the merit of a progeny, the individual in the progeny, and even the replication it belongs to. It was therefore expected that the selection among and within progenies would largely coincide with the BLUP for plants to be selected, which was actually confirmed. Taking as reference the ΣZ, the coincidence was higher than 80 %. When comparing the coincidence of the plants that would be identified by mass or stratified mass selection with those selected by BLUP, the proportion was lower than 40 %.

The results indicate that mass selection was less efficient than BLUP. However, it has been argued that selection by different strategies does not necessarily identify the same progeny/individual, but rather similar individual in terms of performance. In this condition, the efficiency of mass selection strategies would be underestimated. To demonstrate this, the differential of selection (ds) index of the different strategies was estimated. It was observed that the 'ds' was higher in the strategy of mass selection. It could be argued, on this basis, that mass selection was more efficient. The shrinkage effect of BLUP should however be emphasized, which reduces the extreme values and consequently the ds estimate. Moreover, to calculate the gain with selection (GS), the 'ds' must be multiplied by the heritability (h^2). In the case of the BLUP, $h^2 = 1.0$, so, the estimated GS would be the ds estimate. In the selection among and within families on the other hand, the h^2 among and h^2 within should be weighted. Especially the latter should be lower. In the case of mass selection, GS = ds x hI^2, hI^2 is the heritability for selection at the plant level, a value that is usually not high. It can be inferred that the gain expected from mass selection and BLUP would be similar if h2 was 71.6 %, i.e., 2.6 /3.63 = 0.7162. Although this hI^2 estimate was not obtained, it would be difficult reach this value. However, the efficiency of mass selection should not be assumed as only 40 % of BLUP in view of the coincidence of the selected plants.

Coincidence (in %) of the plants selected in the $F_{2:4}$ generation by different selection strategies compared to BLUP considering ΣZ.

Strategy	Best	Worst	ds[1/]
Mass	44	36	3.63
Stratified mass	42	39	3.50
Among and within progencies	83	84	3.14
Blup	100	100	2.60

Multilines as Strategy to Increase Durability of a Cultivar

Most pathogens that infect cultivated plants have several races and new ones often appear. Therefore, the cultivars with resistance to pathogens quickly become susceptible and eventually the cultivar is withdrawn from the market. An option for a more durable resistance is the pyramidization of genes. But although rather time-consuming, pyramidization is no guarantee against the appear of a new race that can break the resistance. Another option, widely acclaimed in the past, are multilines. To evaluate the effectiveness of multilines, Botelho evaluated seven lines of carioca bean separately and in combination or in a multiline. The lines had different resistance alleles to Colletotrichum lindemuthianum, the causal agent of anthracnose. The experiments were conducted in two generation. When the third trifoliate appeared, the plants were inoculated with a mixture of five races of the pathogen.

The estimated of area under the disease progress curve (AUDPC) was greatest for the most susceptible lines (CI - 107 and Carioca). The most significant result, especially in 2007, was that the performance of the multiline was the same as of the resistant lines, bearing in mind that 2/7 of the mixture consisted of very susceptible lines. Apparently, the resistant plants in the mixture acted as a physical barrier against the spread of inoculum, limiting the amount of inoculum reaching the sensitive components of the mixture. The inoculum was also diluted due to its decreased production in resistant plants, and less infective inoculum on susceptible hosts due to the deposition of some of the spores on resistant plants.

For yield, the multiline performed very similar to the lines with best performance. Again this result is very significant, because the two lines Carioca and CI-107 were part of the mixture, with low average grain yields, mainly due to the pathogen attack. Therefore, it can be concluded that the multiline increased stability.

Lines of carioca bean type with the respective number of the race of C. lindemuthianum to which it is resistant, estimate of the area under the disease progression curve (AUDPC) and grain yield (g plot-1) of these lines and the multiline.

Line	Pathogen race	AUDPC		Grain yield ($g.plot_{-1}$)	
Carioca	Susceptible[1/]	35.00 a[2/]	39.25 a	306.67 c	121.67 c
Talisma	89	28.42 b	27.83 b	723.33 b	270.00 b
RC-1-8	337	14.42 c	13.33 d	1043.33 a	276.67 b
MA-11-8	81	11.67 c	12.00 d	1030.33 a	431.67 a
MA-11-16	87	16.17 c	18.75 d	1058.33 a	298.33 b
MA-11-22	65	7.50 c	7.00 d	1108.33 a	275.00 b
CI-107	Susceptible	36.67 a	28.75 b	301.67 c	73.33 c
Multilinha	-	10.83 c	19.17 c	906.67 b	283.33 b

For selection of a new variety from a segregating population after hybridization the methods followed in self-pollinated crop are of three types: 1. Pedigree method, 2. Bulk method, 3. Back cross method.

Pedigree Method

Individual plant progeny is selected from F_2 and subsequent generations, and their progenies are tested. During this process the record of parents as well as off-springs is kept, for which it is known as pedigree method.

The pedigree is defined as the description of the ancestors of an individual and it is generally helpful in finding out the amount of relatedness among two individuals, i.e., whether they are related by common parent in their descent ancestry or not.

Procedure

First Year

The hybridization is done among two selected parents, after emasculation one become female parent and another male parent. After seed set and maturation, the F_1 seeds are harvested separately from each plant individually. On the basis of choice of parents, the type of cross will be of two types — it will be simple cross or complex cross.

Second Year

Seventh Year: Individual plant progenies of F_6 generation are planted in multi-row plots and evaluated visually. Progenies harvested in bulk since they become homozygous. The segregating progenies may be discarded and the preliminary yield trial may be done for the progenies which are reasonably homozygous and have enough seeds.

Eighth Year: Preliminary yield trial with three or more replications is conducted to identify few superior lines. The progenies are evaluated for plant height, lodging, disease resistance, flowering time, maturity time, etc. Quality test is done to serve as an additional basis for selection.

Ninth to Tenth or Thirteenth Year: The superior lines are tested in replicated yield trials at several locations. The line which is superior to the best commercial variety may be released as new variety.

Eleventh or Fourteenth Year: The selected strain should get multiplied to release as a new variety. Breeder has the responsibility to supply the seeds to the state seeds corporation for production and marketing of the seeds.

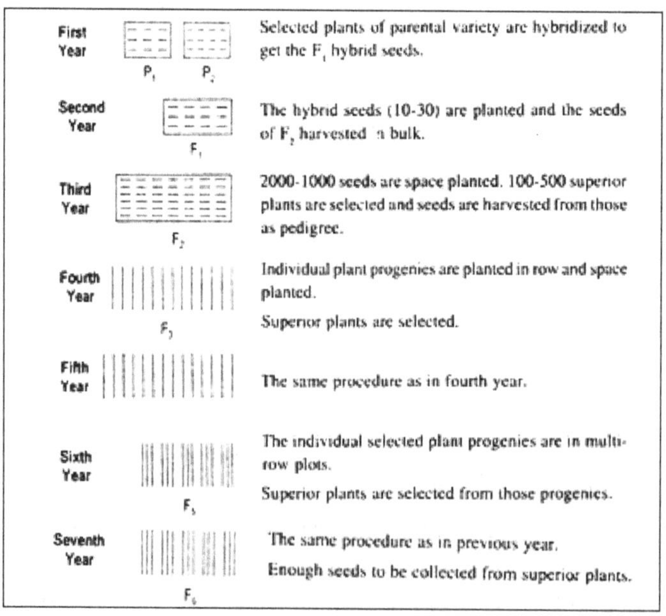

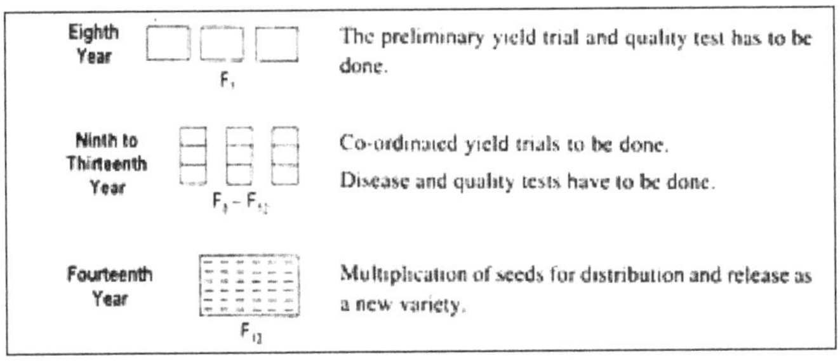

Merits

- This method is most useful as transgressive segregation for yield and other quantitative characters may be recovered in addition to improvement of specific characters.
- This method is well suited for improvement of characters which can be easily identified and simply inherited.
- Through the maintenance of pedigree record the breeder may be able to obtain the information about inheritance of characters.
- Plants or progenies with weaker and visible defects are eliminated at an early stage in the breeding programme.
- This method gives maximum importance on the breeder to use his/her skill and judgement about the selection of plants and progenies.
- This method takes less time than bulk method to release a new variety.

Demerits

- The success of the method is mainly dependent on the skill of the breeder.
- To keep the individual pedigree record is laborious and time consuming, it may be the limiting factor for large breeding programme.
- Selection of large number of progenies in every generation is also laborious and time consuming.
- In F_2 and F_3, the selection for yield is not effective. If sufficient number of progenies is not retained, valuable genotypes may be lost in early segregating generations.

Achievements

Pedigree method is useful in selection of new superior recombinant types from a hybridization programme. This method is suitable for improving specific characteristics, such as disease resistance, plant height, maturity time, etc. as well as yield and quality characters.

Many improved varieties have been developed through pedigree method in many crops like wheat, rice, barley, pulses, oil seeds, cotton, tobacco, jowar, vegetables, etc.

Bulk Method

After hybridization programme when the F_2 and subsequent generations are harvested in mass or as bulk to raise the next generation, it is called bulk method. At the end of bulking period, individual plants are selected and evaluated in the same manner of pedigree method. This method is applicable to any self-pollinating crop like cereals, legumes and oil seeds.

Three different advantages may be achieved from this method:

- Isolation of homozygous lines,
- Waiting for selection by environmental disaster,
- The long period of bulking may be helpful for natural selection to change the composition of population.

Procedure

- First Year: Hybridization is done between the selected parents with the desirable attributes. The cross may be simple or complex depending upon the number of selected parents.
- Second Year: The F_1 generation seeds are space planted and seeds from these plants are harvested in bulk.
- Third Year to Seventh Year: Starting from F_2 generation onwards the seeds are planted at commercial seed rate and spacing. The seeds of subsequent generations are harvested in bulk. Many environmental factors, disease outbreak, etc., may select out the particular genotype from this bulk population. The population size in each generation should be large always, 30,000 – 50,000 plants in each generation.
- Eighth Year: F_7 generation seeds are space planted and superior phenotypes are selected and harvested separately. The selection is made on disease reaction, grain character, etc.
- Ninth Year: Individual plant progenies of F_8 generation are grown in single or multi-row plots. Weak or inferior progenies are rejected, only 100-300 plant progenies are selected which are superior as well as showing no segregation.
- Tenth Year: Preliminary yield trial is conducted with F_9 generation with standard commercial variety as check. The undesirable progenies are rejected.

- Eleventh To Fifteenth Year: Replicated yield trial at several locations is carried out using standard commercial variety. The lines are evaluated for some important characteristics in addition to yield, disease resistance, etc.; superior lines are released as new variety.
- Sixteenth Year: The seeds are multiplied and distributed to the farmer.

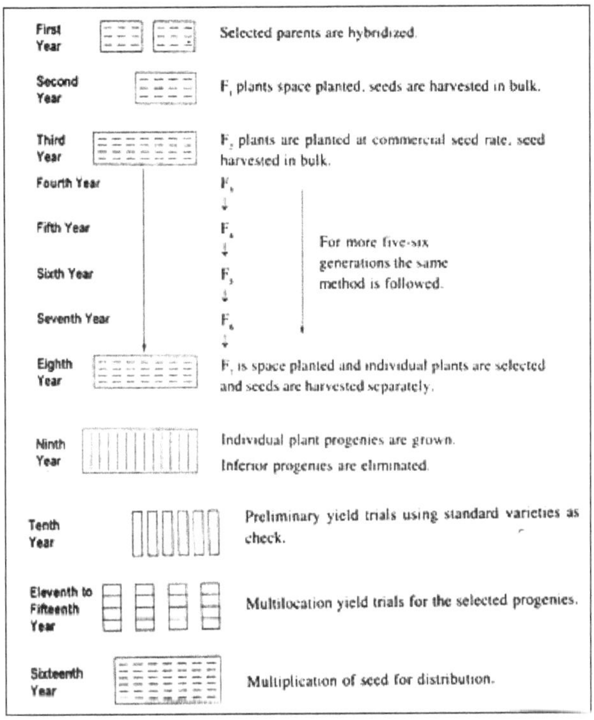

Merits

- This method is simple, convenient and less expensive.
- The natural disease epiphytotics, winter killing, etc., eliminate the undesirable types and increase the frequency of desirable types, which is more helpful for isolation of desirable types.
- As natural selection acts here, so the progenies selected from long term bulk are far superior than those selected from F_2 of short term bulk.
- Less attention and labour is needed for this method.
- Since large populations are grown, the transgressive segregants are more likely to appear and there is greater chance of isolation of transgressive segregants in this method.
- Artificial selection may be practiced to increase the frequency of desirable types.
- Survivability of any gene or a particular genotype can be studied in this method.

Demerits

- This method takes a long time to release a new variety, as the method of natural selection becomes operative only after F_8 or F_{10}.
- The short term bulk method is useful for isolation of homozygous lines but the effect of natural selection has little effect on the genetic composition of the population.
- Breeder has little opportunity to show his skill in this method, practically selection is mainly based on natural selection.
- Inheritance of characters is not individually maintained as in pedigree method.
- At the end of bulking period a large number of progenies are being selected.

Achievements

This method has been used to a limited extent due to lack of popularity. In U.S.A the barley varieties like Arivat, Beecher, Glacier and Gem have been originated from a bulk population derived from cross Atlas x Vaughn, the bulk was maintained in this case for 7 generations.

As this method requires more time to get the effect of natural selection for developing a new variety and the lack of using any skill of breeder for selection of superior types, this method is less used by the breeder for crop improvement.

Genetic Consequences of Repeated Back-crossing

Repeated back-crossing results in decreasing the frequency of heterozygosity by 50% in each subsequent generation and rapid increase in homozygote, upto 98% by sixth back cross progeny, will be achieved. The genotype of the back-cross progeny becomes increasingly similar to that of the recurrent parent.

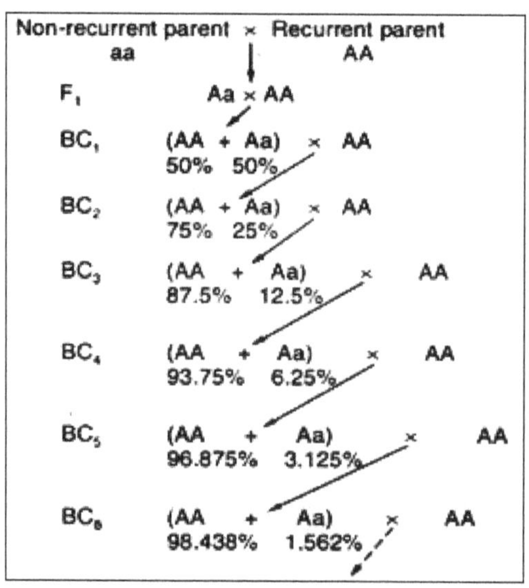

Procedure

The plan of back-cross method differs on the gene which is to be transferred, whether it is dominant or recessive.

Transfer of Dominant Gene

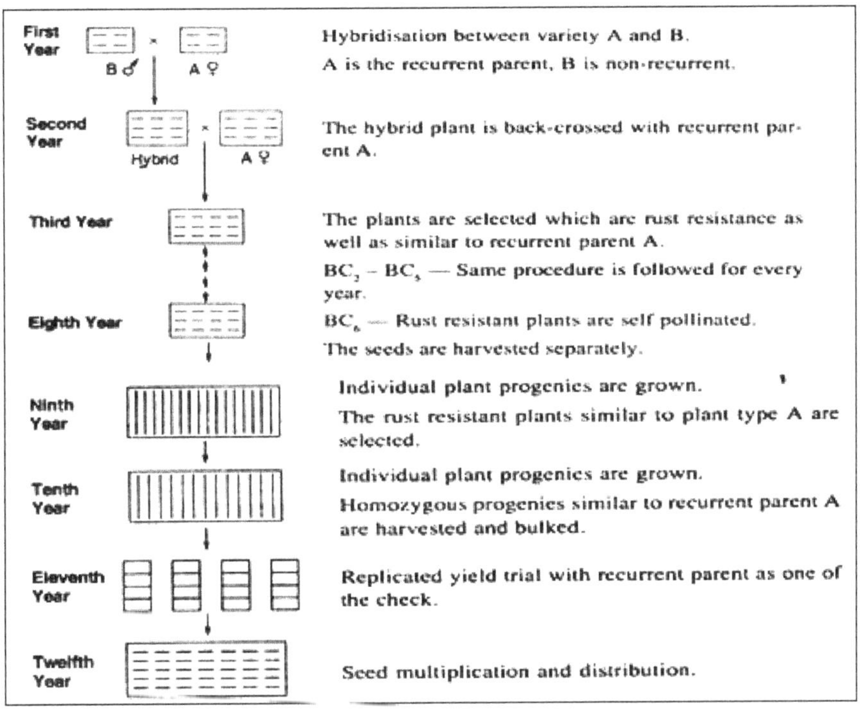

Selection of Parents: Variety A — Recurrent parent (well adapted, high yielding); Variety B — Non-recurrent parent, dominant gene controlled (stem rust resistant).

Hybridization (First Year): The recurrent parent (variety A) should be used as female parent and the non-recurrent parent (variety B) as male parent.

F_1 Generation (Second Year): F_1 plants are back-crossed to variety A, in this generation all the plants are heterozygous for rust resistant character.

BC_1-Generation (Third Year): 50% of the BC_1 plants which are resistant are selected and back-crossed to the recurrent parent, i.e., variety A.

BC_2-BC_5 Generation (Fourth-Seventh Year): In each back-cross generation, segregation would occur for rust resistance. Rust resistant plants are selected and back-crossed to recurrent parent, variety A.

BC_6-Generation (Eighth Year): The plants of this generation will have 98.4% genes from variety A. Rust resistant plants are selected and selfed. Their seeds are harvested separately.

BC_6-F_2 Generation (Ninth Year): Individual plant progenies of selfed plants are grown separately. The rust resistant plants are selected and the seeds are harvested separately.

BC_6-F_3 Generation (Tenth Year): Individual plant progenies are grown. The plants homozygous for rust resistance and similar to the plant type variety A are harvested in bulk.

Yield Test (Eleventh Year): The new variety is tested in replicated yield trial along with the variety A as a check, as the new variety will be same in performance with variety A only with rust resistant character.

Twelfth Year: Seed multiplication is done and distributed to the farmers.

Transfer of Recessive Gene

Procedure for transferring a recessive gene will be different from that of dominant gene, as the recessive gene will be expressed only in homozygous condition. So the selection for that recessive gene requires the F_2 generation, i.e., selfing is needed after every two back-crosses, and testing for the presence of that character has to be done in that generation.

Selection of Parents

The parents are selected in this programme as the previous one, only difference is that the rust resistance is a recessive gene controlled character:

Hybridization (First Year): The recurrent parent is used as female parent and hybridized with the rust resistant donor plant.

F_1 Generation (Second Year): The F_1 plants won't be rust resistant, these will be back-crossed to the recurrent parent.

BC_1 Generation (Third Year): Here also no rust resistant plants will appear so no need to do any rust susceptibility test. All the plants are selfed.

BC_1-F_2 Generation (Fourth Year): The plants are tested for rust resistance character. These plants are selected and back-crossed with recurrent parent.

BC_2 Generation (Fifth Year): There is no rust resistance test, the plants are selected only on resemblances with recurrent parent A, and again back-crossed with recurrent parent.

BC_3 Generation (Sixth Year):Again the plants are self-pollinated, but without any selection for resistance character.

BC_3-F_2 Generation (Seventh Year):The selection for rust resistance is done in this generation. The rust resistant plants having resemblance with variety A are selected and back-crossed.

BC_4 Generation (Eighth Year):Same as BC_2 generation.

BC_5 Generation (Ninth Year):Same as BC_3 generation.

BC_5-F_2 Generation (Tenth Year): Plants are selected against rust resistance and also the characters of variety A. The selfed seeds from the selected plants are harvested separately.

BC_5-F_3 Generation (Eleventh Year): Individual plant progenies are grown and subjected to rust resistance test. Rigid selection is done for resistance to stem rust and the characteristic of variety A. Seeds from selected plants are mixed together to constitute the new variety.

Yield Test (Twelfth Year)

The variety is tested in replicated yield trial along with the variety A as check. The new variety should be same in yield performance except bearing the new character introduced.

Thirteenth Year

Seed multiplication and distribution to the farmers are carried out.

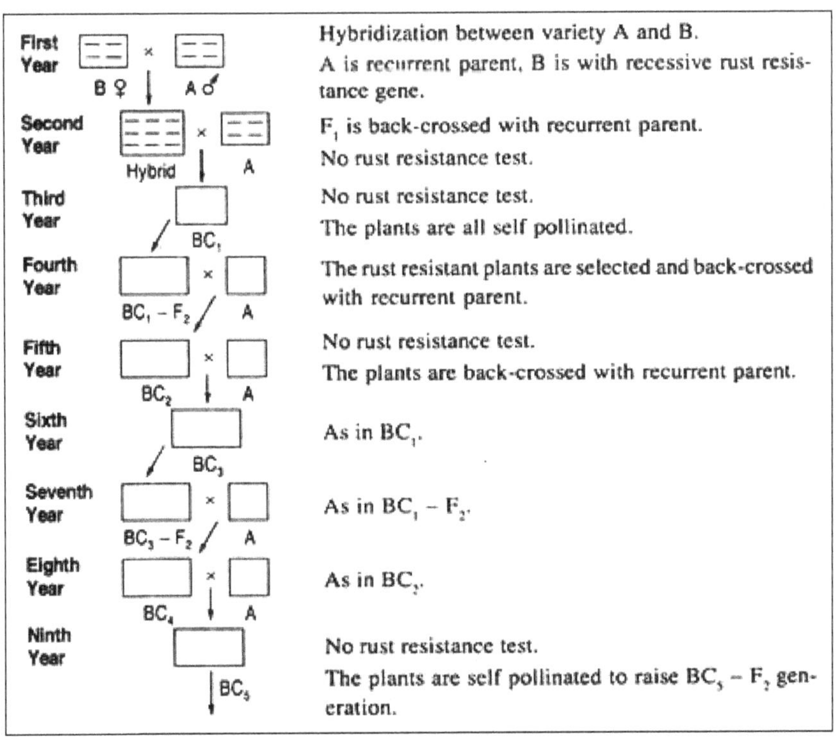

Comparison between the Transfer of Dominant and Recessive Gene during Back- cross Method

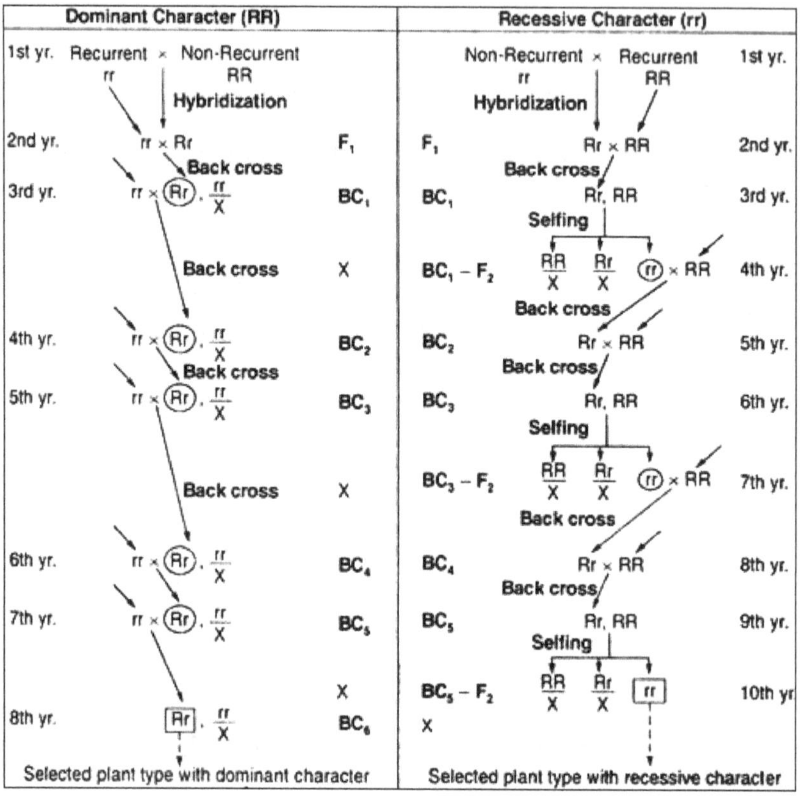

Transfer of Quantitative Character

Quantitative characters are governed by polygenes such as grain size, plant height, maturity time, etc. As during transfer intensity of characters will be lost, so the non-recurrent parent must have the characters more intense from than it is desired in the new variety. For transferring this type of character, each back-cross generation should be selfed and F_2 generation is grown which should have large population.

Then the rigid selection is done for the character to be transferred. The system of alternate back-crossing and selfing goes on till the desired results are obtained. At the end of back-cross programme, $F_2$is handled like pedigree method. The progenies which are more likely to recurrent parent bearing the desirable trait are selected and mixed to make up the new variety.

Transfer of Two or more Characters in a Single Parent

For simultaneous transfer of many characters in a single back-crossing programme, the characters from different non-recurrent parents are brought together by multiple crossing and then back-crossing to recurrent parent is performed. Larger back-cross progenies would be needed than in case of transfer of a single character, and also it may take long time as in each step all the desired characters may not get expressed.

The multiple characters can be transferred in a stepwise method. The recurrent parent is first improved for one character then it is used as recurrent parent for introduction of another new character. Thus in stepwise manner more than one characters can be introduced into a single recurrent parent.

Applications

Generally the.back-cross method is used for transferring disease resistance character to a good and well adapted variety. Other quantitative characters can also be transferred through back-cross method in both cross- and self-pollinated crop.

Inter-varietal Transfer of Simply Inherited Characters

The characters like disease resistance, seed colour, plant height, etc., which are controlled by one or two major genes are most suited for transfer through back-cross method from one variety to another of the same species. The successful transfer depends on the minimum linkage between desirable and undesirable trait.

Inter-varietal Transfer of Quantitative Characters

Grain characters like seed size and shape, earliness, plant height-all can be transferred from one variety to another with the criteria of high heritability.

Interspecific Transfer of Simply Inherited Characters

Mainly the character like disease resistance can be transferred from related species to cultivated species. This specially requires the relatedness between the species where the chromosomes can pair during meiosis. The transfer will be unsuccessful if the genetic environment of recurrent parent is not suitable for functioning of the gene of desirable character from donor parent.

Transfer of Cytoplasm

In case of transfer of male sterile character from one parent to another requires back-cross method. The variety or species from which the cytoplasm is to be transferred is used as female parent. The recurrent parent should be the male parent. After 6-8 back-crosses, the cytoplasm will be of the donor parent with the genotype of recurrent parent.

Transgressive Segregation

Modification of back-cross method will produce transgressive segregants. Few (1 to 3) back-crosses with F_1 allow much heterozygosity to appear, or two or more recurrent parents may be used in back- cross programme to accumulate genes from them into the back-cross progeny. Such kind of modification would produce new variety which won't be like the recurrent parent.

Production of Isogenic Lines

Back-cross method is useful for production of isogenic lines, i.e., the lines of a crop which are identical in their genotype except for one gene. These isogenic lines are useful for studying the effect of individual genes.

Merits

- This method does not change the genotype of the popular established variety, only it helps a single desirable character to be transferred in the existing variety.
- As the recurrent parent is an established variety, so it is not necessary to test the yield performance which ultimately saves five years-time, as well as expenses.
- In case of short duration plant, 2-3 generations can be raised within a year as the selection is based on inheritance of a particular character, not on performance. So this method drastically reduces the time period required to develop a new variety.
- Back-cross method requires smaller populations than the pedigree method.
- Defects of an established variety can be removed using this method, may be only by introducing a single character.
- Interspecific gene transfer only can be done through this method.
- Transgressive segregation can be obtained in case of quantitative characters by modified back-cross method.
- This method is very much useful for cytoplasmic gene transfer to the recurrent parent, i.e., new variety will bear the cytoplasm of donor parent and genotype of recurrent parent.

Demerits

- The new variety is not superior in performance than the existing variety except the introduction of single character.
- During transfer of such kind of single desirable gene, sometimes some undesirable genes may get transferred.
- For introduction of more than one gene controlled character, multiple crossing programme is necessary which is often difficult, time taking and costly.

- Once a recurrent parent taken in a back-cross method may get replaced by another superior variety of high yielding ability.

Achievements

Back-cross method is very much useful for transferring of simply heritable character like disease resistance to well adapted popular local variety.

For interspecific transfer of genes, back-cross method is widely used for crop improvement. Cultivated sugarcane (Saccharum officincirum) is susceptible to pests and disease, crossed with S. spontanium which is resistant. This brings resistance but with undesirable characters like more fibre, low sugar, thin stem, etc. By back-crossing with noble cane these undesired characters are removed.

In cotton, the hybridization between Gossypium hirsutum and G. arboreum yielded highly sterile F hybrid, few tetraploid seeds were obtained. These plants were then back- crossed with G. hirsutum and two varieties have been selected from the back-cross progeny which are now being widely cultivated.

References

- Mass-selection-features-types-methods-crop-improvement-botany, plant-breeding-2, breeding-methods, mass-selection-14285: botanylibrary.com, Retrieved 16 April, 2020
- Improvement, crops, hybridization-method-of-crop-improvement-17701: biologydiscussion.com, Retrieved 20 April, 2020
- Breeding-self-pollinated-plants-262596217: researchgate.net, Retrieved 10 June, 2020
- Pure-line-selection-in-crops-meaning-and-theory-plant breeding-botany, plant-breeding-2, breeding-methods, pure-line-selection-13968: botanylibrary.com, Retrieved 14 March, 2020
- Pedigree-method-and-back-cross-method-breeding, breeding-60826: biologydiscussion.com, Retrieved 09 March, 2020

Breeding for Stress Tolerance

4

- Breeding for Drought Stress Tolerance
- Breeding for Heat Stress Tolerance
- Breeding for Mechanical Weed Control
- Breeding for Waterlogging Tolerance
- Engineering Cold Stress Tolerance in Crop Plants
- Breeding for Salinity Tolerance in Plants

Plant breeding is concerned with management of various stress tolerance such as drought stress tolerance, heat stress tolerance, waterlogging tolerance, cold stress tolerance, salinity tolerance, etc. This chapter discusses breeding for stress tolerance for a thorough understanding of the subject.

Historically, farmers have used a trial and error approach in selecting the crops that they grow, choosing particular crops because they perform well under their climatic and field conditions. Plant breeders have expended considerable effort to develop high-yielding cultivars of those crops which are adapted to relatively optimal environments. Through these endeavors the efficiency of food production has improved steadily. But as the demand for food and fiber increases, the impact of fluctuating environments becomes increasingly important. It is the sensitivity of our crop plants to environmental stresses which limits their climatic distribution and time of year for production. In many areas of the world, the environmental milieu is marginal, or totally unsuited for economic production of crop plants. Even in areas normally suited to these crops, sensitivity to sporadic environmental insults at critical periods during plant growth and development also results in losses. The intense one (or two)-crop agriculture practiced in certain areas of many developed countries is particularly vulnerable to short-term fluctuations in the environment; the crops used having been selected for genetic uniformity and developed for optimal conditions. Breeding crops for high yield potential will undoubtedly continue as a preeminent breeding objective. The importance of yield stability across different environmental regimes, however, is gaining interest.

The major climatic variables in productivity are temperature and water availability. It has been estimated that 60 to 80% of the seasonal variation in crop productivity is due to weather fluctuations, which includes growing conditions having extremes in temperature and water availability. Temporary over-irrigation may cause localized water-logging of the soil. The resultant reduction in oxygen levels may be sufficient to reduce plant growth significantly (3). Flood irrigation of fields as practiced in many regions of the world results in temporary periods of anaerobic stress. Water seepage from irrigation ditches may also cause soil to become waterlogged and substantially reduce crop yields (4). Other environmental conditions may also tend to exacerbate this problem. In temperate regions where prolonged winters are encountered, ice crusts may seal the soil surface and affect gas diffusion to the impediment of root function of perennial winter crops. Of concern also are suboptimal soil conditions with either mineral deficiencies or toxicities; or soils with pH extremes. Soils with biologically counterproductive or stressful mineral properties, including alkalinity or salinity, have been estimated to occupy nearly 30% of the world's land area (5). Several additional stresses, more localized and therefore less important to crop productivity on a world-wide basis, include air movement (wind), air pollutants, light quantity (cloud cover or fog), humidity and hail.

Increased production must come either from higher yields on existing cropland or expansion to production on less arable land. Yield plateaus have been suggested for several major crops and the feasibility of even greater production through increased use of fertilizer, water and pesticides is waning. Reliance upon production at the maximal end of the yield spectrum intensifies further the impact of uncontrollable fluctuations in the environment. Expansion of crop production onto marginal land areas presents numerous stress constraints to economic production.

Future limitation of energy-expensive inputs may lower the quality of environments presently having high productivity. Any significant increase in productivity is dependent upon our ability to adapt crops to cope better with climatic fluctuations and less than optimal soil conditions.

Breeding for Drought Stress Tolerance

Breeding for drought resistance is the process of breeding plants with the goal of reducing the impact of dehydration on plant growth.

Dehydration Stress

Crop Plants

In nature or crop fields, water is often the most limiting factor for plant growth. If plants do not receive adequate rainfall or irrigation, the resulting dehydration stress can reduce growth more than all other environmental stresses combined.

Drought can be defined as the absence of rainfall or irrigation for a period of time sufficient to deplete soil moisture and cause dehydration in plant tissues. Dehydration stress results when water loss from the plant exceeds the ability of the plant's roots to absorb water and when the plant's water content is reduced enough to interfere with normal plant processes.

Global Phenomenon

About 15 million km2 of the land surface is covered by crop-land, and about 16% of this area is equipped for irrigation. Thus, in many parts of the world, including the United States, plants may frequently encounter dehydration stress. Rainfall is very seasonal and periodic drought occurs regularly. The effect of drought is more prominent in sandy soils with low water holding capacity. On such soils some plants may experience dehydration stress after only a few days without water.

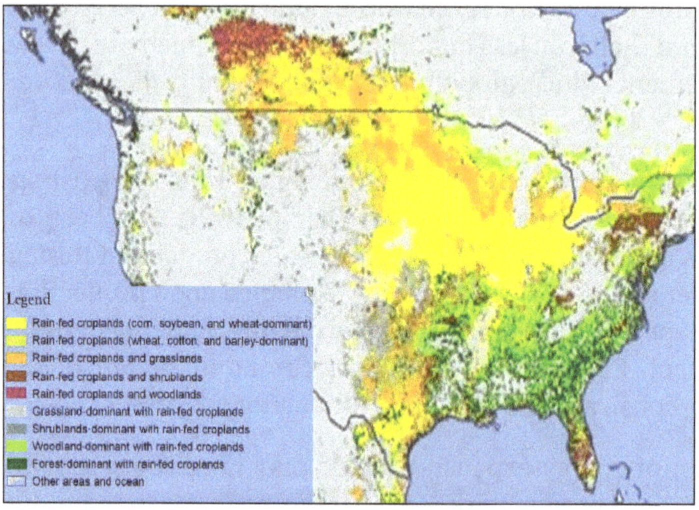

Rain fed areas of USA.

During the 20th century, the rate of increase in 'blue' water withdrawal (from rivers, lakes, and aquifers) for irrigation and other purposes was higher than the growth rate of the world population. Country-wise maps of irrigated areas are available.

Future Challenges to Crop Production

Soil moisture deficit is a significant challenge to the future of crop production. Severe drought in parts of the U.S., Australia, and Africa in recent years drastically reduced crop yields and disrupted regional economies. Even in average years, however, many agricultural regions, including the U.S. Great Plains, suffer from chronic soil moisture deficits. Cereal crops typically attain only about 25% of their potential yield due to the effects of environmental stress, with dehydration stress the most important cause. Two major trends will likely increase the frequency and severity of soil moisture deficits:

- Climate change: Higher temperatures are likely to increase crop water use due to increased transpiration. A warmer atmosphere will also speed up melting of mountain snow pack, resulting in less water available for irrigation. More extreme weather patterns will increase the frequency of drought in some regions.
- Limited water supplies: Increased demand from municipal and industrial users will further reduce the amount of water available for irrigated crops.

Although changes in tillage and irrigation practices can improve production by conserving water, enhancing the genetic tolerance of crops to drought stress is considered an essential strategy for addressing moisture deficits.

Plant Physiology

A plant responds to a lack of water by halting growth and reducing photosynthesis and other plant processes in order to reduce water use. As water loss progresses, leaves of some species may appear to change colour – usually to blue-green. Foliage begins to wilt and, if the plant is not irrigated, leaves will fall off and the plant will eventually die. Soil moisture deficit lowers the water potential of a plant's root and, upon extended exposure, abscisic acid is accumulated and eventually stomatal closure occurs. This reduces a plant's leaf relative water content.

The time required for dehydration stress to occur depends on the water-holding capacity of the soil, environmental conditions, stage of plant growth, and plant species. Plants growing in sandy soils with low water-holding capacity are more susceptible to dehydration stress than plants growing in clay soils. A limited root system will accelerate the rate at which dehydration stress develops. A plant's root system may be limited by the presence of competing root systems from neighbouring plants, by site conditions such as compacted soils or high water tables, or by container size (if growing in a container). A plant with a large mass of leaves in relation to the root system is prone to drought stress as the leaves may lose water faster than the roots can supply it. Newly planted plants and poorly established plants may be especially susceptible to dehydration stress because of the limited root system or the large mass of stems and leaves in comparison to roots.

Other Stress Factors

Aside from the moisture content of the soil, environmental conditions of high light intensity, high temperature, low relative humidity and high wind speed will significantly increase plant water loss. The prior environment of a plant also can influence the development of dehydration stress. A plant that has been exposed to dehydration stress (hardened) previously and has recovered may become more drought resistant. Also, a plant that was well-watered prior to being water-limited will usually survive a period of drought better than a continuously dehydration-stressed plant.

Mechanisms of Drought Resistance

The degree of resistance to drought depends upon individual crops. Generally three strategies can help a crop to mitigate the effect of dehydration stress:

Avoidance

If the plant shows dehydration avoidance, the environmental factor is excluded from the plant tissues by reducing water loss ("water savers", e.g. by thick leaf epicuticular wax, leaf rolling, leaf posture) or maintaining water uptake ("water spenders", e.g. by growing deeper roots). Dehydration avoidance is desirable in modern agriculture, where drought resistance requires the maintenance of economically viable plant production under dehydration stress. The role of dehydration avoidance is maintaining water supply and sustaining leaf hydration and turgidity with the purpose of maintaining stomatal opening and transpiration as long as possible under water deficit. This is essential for leaf gas exchange, photosynthesis and plant production through carbon assimilation.

Tolerance

If the plant shows dehydration tolerance, the environmental factor enters the plant tissues but the tissues survive, by e.g. maintaining turgor and osmotic adjustment.

Escape

Dehydration escape involves e.g. early maturing or seed dormancy, where the plant uses previous optimal conditions to develop vigor. Dehydration recovery refers to some plant species being able to recuperate after brief drought periods.

A proper timing of life-cycle, resulting in the completion of the most sensitive developmental stages while water is abundant, is considered to be a dehydration escape strategy. Avoiding dehydration stress with a root system capable of extracting water from deep soil layers, or by reducing evapotranspiration without affecting yields, is considered as dehydration avoidance. Mechanisms such as osmotic adjustment (OA) whereby a plant maintains cell turgor pressure under reduced soil water potential are categorised as dehydration tolerance mechanisms. Dehydration avoidance mechanisms can be expressed even in the absence of stress and are then considered constitutive. Dehydration tolerance mechanisms are the result of a response triggered by dehydration stress itself and are therefore considered adaptive. When the stress is terminal and predictable, dehydration escape through the use of shorter duration varieties is often the preferable method of improving yield potential. Dehydration avoidance and tolerance mechanisms are required in situations where the timing of drought is mostly unpredictable.

Drought resistance mechanisms are genetically controlled and genes or QTL responsible for drought resistance have been discovered in several crops which opens avenue for molecular breeding for drought resistance.

Drought Resistance Traits

Resistance to drought is a quantitative trait, with a complex phenotype, often confounded by plant phenology. Breeding for drought resistance is further complicated since several types of abiotic stress, such as high temperatures, high irradiance, and nutrient toxicities or deficiencies can challenge crop plants simultaneously.

Osmotic Adjustment

When a plant is exposed to water deficit, it may accumulate a variety of osmotically active compounds such as amino acids and sugars, resulting in a lowering of the osmotic potential. Examples of amino acids that may be up-regulated are proline and glycine betaine. This is termed osmotic adjustment and enables the plant to take up water, maintain turgor and survive longer.

Cell Membrane Stability

The ability to survive dehydration is influenced by a cell's ability to survive at reduced water content. This can be considered complementary to OA because both traits will help maintain leaf growth (or prevent leaf death) during water deficit. Crop varieties differ in dehydration tolerance and an important factor for such differences is the capacity of the cell membrane to prevent electrolyte leakage at decreasing water content, or "cell membrane stability (CMS)". The maintenance of membrane function is assumed to mean that cell activity is also maintained. Measurements of CMS have been used in different crops and are known to be correlated with yields under high temperature and possibly under dehydration stress.

Epicuticular Wax

In sorghum (Sorghum bicolor L. Moench), drought resistance is a trait that is highly correlated with the thickness of the epicuticular wax layer. Experiments have demonstrated that rice varieties with a thick cuticle layer retain their leaf turgor for longer periods of time after the onset of a water-stress.

Partitioning and Stem Reserve Mobilisation

As photosynthesis becomes inhibited by dehydration, the grain filling process becomes increasingly reliant on stem reserve utilisation. Numerous studies have reported that stem reserve mobilisation capacity is related to yield under dehydration stress in wheat. In rice, a few studies also indicated that this mechanism maintains grain yield under dehydration stress at the grain filling stage. This dehydration tolerance mechanism is stimulated by a decrease in gibberellic acid concentration and an increase in abscisic acid concentration.

Manipulation and Stability of Flowering Processes

Seedling Drought Resistance Traits

For emergence from deep sowing (to exploit dry upper soil), this is practised to help seedlings reach the receding moisture profile, and to avoid high soil surface temperatures which inhibit germination. Screening at these stage provides practical advantages, specially when managing large amount of germ-plasms.

Breeding for Heat Stress Tolerance

heat stress is one of such factor that reduces the production and quality significantly. So breeding against heat is a very important criterion for breeding for current as well as future environments produced by global climate change (e.g. global warming).

Breeding for Heat Stress Tolerance in Plants

Heat stress due to increased temperature is a very important problem globally. Occasional or prolonged high temperatures cause different morpho-anatomical, physiological and biochemical changes in plants. The ultimate effect is on plant growth as well as development and reduced yield and quality. Breeding for heat stress tolerance can be mitigated by breeding plant varieties that have improved levels of thermo-tolerance using different conventional or advanced genetic tools. Marker assisted selection techniques for breeding are highly useful. Recently 41 polymorphic SSR markers has been identified between a heat tolerant rice variety 'N22' and heat susceptible-high yielding variety 'Uma' for the development of new 'high yielding-heat tolerant' rice varieties.

What is Heat Stress Tolerance?

Heat stress is defined as increased temperature level sufficient to cause irreversible damage to plant growth and development. Generally a temperature rise, above usually 10 to 15 °C above ambient, can be considered heat shock or heat stress. Heat tolerance is broadly defined as the ability of the plant tolerate heat – means that grow and produce economic yield under high temperatures.

Significance: Current and Future

Heat stress is a serious threat to crop production globally. Global warming is particularly consequence of increased level of green house gases such as CO_2, methane, chlorofluorocarbons and nitrous oxides. The Intergovernmental Panel on Climatic Change (IPCC) has predicted a rise of 0.3 °C per decade reaching to approximately 1 and 3 °C above the present value by 2025 and 2100 AD, respectively.

Physiological Consequence of Heat Stress

At very high temperatures cause severe cellular injury and cell death may occur within short time, thus leading to a catastrophic collapse of cellular organization. However, under moderately high temperatures, the injury can only occur after longer exposure to such a temperature however the plant efficiency can be severely affected. High temperature directly affect injuries such as protein denaturation and aggregation, and increased fluidity of membrane lipids. Other indirect or slower heat injuries involve inactivation of enzymes in chloroplast and mitochondria, protein degradation, inhibition of protein synthesis, and loss of membrane integrity. Heat stress associated injuries ultimately lead to starvation, inhibition of growth, reduced ion flux, production of toxic compounds and production of reactive oxygen species (ROS). Immediately after exposure to high temperature stress-related proteins are expressed as stress defense strategy of the cell. Expression of heat shock proteins (HSPs), protein with 10 to 200 kDa, is supposed to be involved in signal transduction during heat stress. In many species it has been demonstrated that HSPs results in improved physiological phenomena such as photosynthesis, assimilate partitioning, water and nutrient use efficiency, and membrane stability.

Studies have found tremendous variation within and between species, thus this will help to breed heat tolerance for future environment. Some of attempts to develop heat-tolerant genotypes are successful.

Traits Associated with Heat Stress Tolerance

Different physiological mechanisms may contribute to heat tolerance in the field—for example, heat tolerant metabolism as indicated by higher photosynthetic rates, stay-green, and membrane thermo-stability, or heat avoidance as indicated by canopy temperature depression. Several physiological and morphological traits have been evaluated for heat tolerance - Canopy temperature, leaf chlorophyll, stay green, leaf conductance, spike number, biomass, and flowering date.

Canopy Temperature Depression

CTD has shown clear association with yield in warm environments shows it association with heat stress tolerance. CTD shows high genetic correlation with yield and high values of proportion of direct response to selection indicating that the trait is heritable and therefore amenable to early generation selection. Since an integrated CTD value can be measured almost instantaneously on scores of plants in a small breeding plot (thus reducing error normally associated with traits measured on individual plants), work has been conducted to evaluate its potential as an indirect selection criterion for genetic gains in yield. CTD is affected by many physiological factors, which makes it a powerful.

Stomatal Conductance

Canopy temperature depression is highly suitable for selecting physiologically superior lines in warm, low relative humidity environments where high evaporative demand leads to leaf cooling of up to 10 °C below ambient temperatures. This permits differences among genotypes to be detected relatively easily using infrared thermometry. However, such differences cannot be detected in high relative humidity environments because the effect of evaporative cooling of leaves is negligible. Nonetheless, leaves maintain their stomata open to permit the uptake of CO_2, and differences in the rate of CO_2 fixation may lead to differences in leaf conductance that can be measured using a porometer. Porometry can be used to screen individual plants. The heritability of stomatal conductance is reasonably high, with reported values typically in the range of 0.5 to 0.8. Plants can be assessed for leaf conductance using a viscous flow porometer that is available on the market (Thermoline and CSIRO, Australia). This instrument can give a relative measure of stomatal conductance in a few seconds, making it possible to identify physiologically superior genotypes from within bulks.

Membrane Thermostability

Although resistance to high temperatures involves several complex tolerance and avoidance mechanisms, the membrane is thought to be a site of primary physiological injury by heat, and measurement of solute leakage from tissue can be used to estimate damage to membranes. Since membrane thermostability is reasonably heritable and shows high genetic correlation with yield.

Chlorophyll Fluorescence

Chlorophyll fluorescence, an indication of the fate of excitation energy in the photosynthetic apparatus, has been used indicator for heat stress tolerance.

Chlorophyll Content and Stay Green

Chlorophyll content and stay green traits have been found to be associated with heat stress tolerance.,. Xu identified three QTLs for chlorophyll content (Chl1, Chl2, and Chl3) (coincided with three stay-green QTL regions (Stg1, Stg2, and Stg3)) were identified in Sorghum. The Stg1 and Stg2 regions also contain the genes for key photosynthetic enzymes, heat shock proteins, and an abscisic acid (ABA) responsive gene.

Photosynthesis

Decliend photosynthesis is suggested as measure of heat stress sensitivity in plants.

Breeding for Mechanical Weed Control

Mechanical weed control is an physical activity that inhibits unwanted plant growth. Mechanical, or manual, weed control techniques manage weed populations through physical methods that remove, injure, kill, or make the growing conditions unfavorable. Some of these methods cause direct damage to the weeds through complete removal or causing a lethal injury. Other techniques may alter the growing environment by eliminating light, increasing the temperature of the soil, or depriving the plant of carbon dioxide or oxygen. Mechanical control techniques can be either selective or non-selective. A selective method has very little impact on non-target plants where as a non-selective method affects the entire area that is being treated. If mechanical control methods are applied at the optimal time and intensity, some weed species may be controlled or even eradicated.

Mechanical Control Methods

Weed Pulling

Pulling methods uproot and remove the weed from the soil. Weed pulling can be used to control some shrubs, tree saplings, and herbaceous plants. Annuals and tap-rooted weeds tend to be very susceptible to pulling. Many species are able to re-sprout from root segments that are left in the soil. Therefore, the effectiveness of this method is dependent on the removal of as much of the root system as possible. Well established perennial weeds are much less effectively controlled because of the difficulty of removing all of the root system and perennating plant parts. Small herbaceous weeds may be pulled by hand but larger plants may require the use of puller tools like the Weed Wrench or the Root Talon. This technique has a little to no impact on neighboring, non-target plants and has a minimal effect on the growing environment. However, pulling is labor-intensive and time consuming making it a more suitable method to use for small weed infestations.

Mowing

Mowing methods cut or shreds the above ground of the weed and can prevent and reduce seed populations as well as restrict the growth of weeds. Mowing can be a very successful control method for many annual weeds. Mowing is the most effective when it is performed before the weeds are able to set seed because it can reduce the number of flower stalks and prevent the spread of more seed. However, the biology of the weed must be considered before mowing. Some weed species may sprout with increased vigor after being mowed. Also, some species are able to re-sprout from stem or root segments that are left behind after mowing. Brush cutting and weed eating are also mowing techniques that reduce the biomass of the weeds. Repeatedly removing biomass causes reduced vigor in many weed species. This method is usually used in combination with other control methods such as burning or herbicide treatments.

Mulching

Mulch is a layer of material that is spread on the ground. Compared with some other methods of weed control, mulch is relatively simple and inexpensive. Mulching smothers the weeds by excluding light and providing a physical barrier to impede their emergence. Mulching is successful with most annual weeds, however, some perennial weeds are not affected. Mulches may be organic or synthetic. Organic mulches consist of plant by products such as: pine straw, wood chips, green waste, compost, leaves, and grass clippings. Synthetic mulches, also known as ground cover fabric, can be made from materials like polyethylene, polypropylene, or polyester. The effectiveness of mulching is mostly dependent on the material used. Organic and synthetic mulches may be used in combination with each other to increase the amount of weeds controlled.

Tillage

Tillage, also known as cultivation, is the turning over of the soil. This method is more often used in agricultural crops. Tillage can be performed on a small scale with tools such as small, hand pushed rotary tillers or on a large scale with tractor mounted plows. Tillage is able to control weeds because when the soil is overturned, the vegetative parts of the plants are damaged and the root systems are exposed causing desiccation. Generally, the younger the weed is, the more readily it can be controlled with tillage. To control mature perennial weeds, repeated tillage is necessary. By continually destroying new growth and damaging the root system, the weed's food stores are depleted until it can no longer resprout. Also, when the soil is overturned, the soil seed bank is disrupted which can cause dormant weed seeds to germinate in the absence of the previous competitors. These new weeds can also be controlled by continued tillage until the soil seed bank is depleted.

Soil Solarization

Soil solarization is a simple method of weed control that is accomplished by covering the soil with a layer of clear or black plastic. The plastic that is covering the ground traps heat energy from the sun and raises the temperature of the soil. Many weed seeds and vegetative propagules are not able to withstand the temperatures and are killed. For this method to be most effective, it should be implemented during the summer months and the soil should be moist. Also, cool season weeds are more susceptible to soil solarization than are warm season weeds. Using black plastic as a cover excludes light which can help to control plants that are growing whereas clear plastic has been shown to produce higher soil temperatures.

Fire

Burning and flaming can be economical and practical methods of weed control if used carefully. For most plants, fire causes the cell walls to rupture when they reach a temperature of 45 °C to 55 °C. Burning is commonly used to control weeds in forests, ditches, and roadsides. Burning can be used to remove accumulated vegetation by destroying the dry, matured plant matter as well as killing the green new growth. Buried weed seeds and plant propagules may also be destroyed during burning, however, dry seeds are much less susceptible to the increased temperature. Flaming is used on a smaller scale and includes the use of a propane torch with a fan tip. Flaming may be used to control weeds along fences and paved areas or places where the soil may be too wet to hoe, dig, or till. Flaming is most effective on young weeds that are less than two inches tall but repeated treatments may control tougher perennial weeds.

Flooding

Flooding is a method of control that requires the area being treated to be saturated at a depth of 15 to 30 cm for a period of 3 to 8 weeks. The saturation of the soil reduces the availability of oxygen to the plant roots thereby killing the weed. This method has been shown to be highly effective in controlling establish perennial weeds and may also suppress annual weeds by reducing the weed seed populations.

Effects of Mechanical Control on the Environment

Mechanical methods of weed control cause physical changes in the immediate environment that may cause positive or negative effects. The suppression of the targeted weeds will open niches in the environment and may also stimulate the growth of other weeds by decreasing their competition and making their environment more favorable. If the niches are not filled by a desirable plant, they will eventually be taken over by another weed. These weed control methods also effect the structure of the soil. The use of mulches can help decrease erosion, decrease water evaporation from the soil, as well as improve the soil structure by increasing the amount of organic matter. Tillage practices can help decrease compaction and aerate the soil. On the other hand, tillage has also been shown to decrease soil moisture, increase soil erosion and runoff, as well as decrease soil microbial populations. Solarization can cause changes in the biological, physical, and chemical properties of the soil. This can cause the soil to be an unfavorable environment for native species which may be beneficial or harmful.

Breeding for Waterlogging Tolerance

Waterlogging is a serious problem in low-laying rainfed areas. Lack of oxygen supply for the plant is the main reason of damage in waterlogging condition, which hampers nutrient and water uptake, as a reason the plant shows wilting. In oxygen-deprived condition plants shift its metabolism to anaerobic from aerobic mode. Plants which can withstand waterlogging condition have mechanisms such as increased availability of soluble sugar, aerenchyma formation, greater activity of glycolytic pathway and fermentation enzymes, and involvement of antioxidant defense mechanism to cope with the oxidative stress induced by waterlogging. Ethylene plays an important role in change of mechanisms of plants in deficiency of oxygen. It was also reported that ethylene induces the genes of enzymes associated with aerenchyma formation, glycolysis and fermentation pathway.

Hypoxia and Anoxia

Roots of the plants are usually in contact with oxygen at a partial pressure which is equivalent to the gaseous atmosphere. Hypoxia, reduction of oxygen below optimum level, is the most common form of stress which occurs during partial submergence of plant due to short-term flooding where the root goes under water and shoots remain in the atmosphere. Anoxia, another kind of water stress where plant goes under water completely, hence complete absence of oxygen as a result of long-term flooding. Microbial flora of the soil may change by long time waterlogging which works in favour of anaerobic microorganisms that use, as alternative, electron acceptors to oxygen. In such condition soil tends to accumulate nitrite as it tends to accumulate more reduced and phytotoxic forms of mineral ions (from nitrate) and ferrous ions (from ferric) and very few number of plants are naturally adapted to grow in this kind of soil.

Ethylene Production and Flooding

Accumulation of ethylene in waterlogged soil and plants occurs at concentration of 10 $cm^3 dm^{-3}$. This accumulation of ethylene occurs in mainly two ways, firstly ethylene diffusion rate from root to water is 10 times slower than in air and, ethylene synthesis is increased in the hypoxic root system and in the aerobic shoot. At first ethylene may be released to the internal channels of aerenchyma and then diffused to the root area. In roots high amount of ACC (1-amino cyclopropane 1-carboxylic acid) synthesis is occurred, which is a precursor of ethylene. ACC to ethylene conversion cannot happen without oxygen; as a result this reaction is blocked in the aerobic root system. Formation steps of ACC do not require oxygen, furthermore, accumulation of ACC rather increased in aerobic condition. As a result the amount of ACC is increased to the shoot. Highest amount of ACC accumulation occurs in the lowest portions of the stem and ethylene is released in the presence of oxygen. In partial submergence rice undergoes through a special pathway initiated by the entrapment of ethylene in the plant cell. As a result ethylene encourages degradation of abscisic acid (ABA) and increases the content of gibberellic acid (GA) and their downstream effects. Remarkably, the rates of stem elongation in deepwater rice varieties may be up to 25 cm/day. This phenomenon of deepwater rice is mainly controlled by three QTLs. Of these, the SNORKEL QTL on chromosome 12 encodes two ethylene responsive factor (ERF) DNA binding proteins, SNORKEL1 (SK1), and SNORKEL2 (SK2), which were absent to nondeepwater rice accessions.

Role of Ethylene in Aerenchyma Formation

Formation of aerenchyma is initiated by ethylene, which is one of many adaptive features of plants at submergence to avoid anaerobiosis by increasing the availability of oxygen. The soft tissues with large intercellular spaces to render low-resistance internal pathways for the exchange of gases between aerobic shoot to the anaerobic root are called Aerenchyma. This feature has been reported in many species like, *Trifolium subterraneum*, wheat, rice, barley, maize, soybea, *Vigna luteola,* and *Carex* spp. Oxygen leaks out into the roots and surrounding soil through the aerenchyma. As a result, a small oxygenated soil environment formed which might create aerobic environment to microorganisms and prevent the development of potentially toxic soil components such as oxides of Iron, Cu, and Mn. Synthesis of ethylene was highly increased in roots under hypoxia condition, simultaneously with the formation of aerenchyma.

Marker Assisted Breeding for Submergence Tolerance

Since in mid 1990s major genes responsible for waterlogging tolerance have been identified and it become easy for the researcher to concentrate on modification or use of those genes to develop new waterlogging tolerant crop varieties like rice, wheat, maize, soybean, barley, and so forth. In the case of rice introgression of *Sub1* gene to specific varieties by marker assisted backcrossing (MAB) for various land types and choice of the farmer and addition of new varieties through genetic engineering became possible.

Table: QTL identification of waterlogging tolerance in different crops.

Species	QTL/chromosome No./genes	Marker used	Method
Rice (Oryza sativa L.)	Sub1 (Ch. 9)	SSR	MAS
Barley (Hordeum vulgare L.)	Qwt4-1	SSR	MAS
Barley (Hordeum vulgare L.)	tfy2.1-1, tfy1.1-2, tfy1.2-1	RFLP	MAS

Barley (Hordeum vulgare L.)	tfy1.1-3, tfsur-2 tfsur-1, tfy1.1-1, tfmas, tfy2.1-2	DArT	MAS
Maize (Zea mays L.)	Ch. 4, 9	SSR	CIM
Maize (Zea mays L.)	Ch. 1–3, 6, 7, and Ch. 10	SSR	CIM
Soybean [Glycine max (L.)Merr.]	Rps	SSR	SMA
Soybean [Glycine max (L.) Merr.]	Rps	SSR	CIM

Identification of QTL Associated with Waterlogging Tolerance

Genetic control of submergence tolerance was unknown until mid-1990s. It was suggested as a quantitative trait. *Sub1* the major QTL responsible for submergence tolerance was found out by molecular mapping on chromosome 9, which contributes almost 70% of phenotypic variation in tolerance. Furthermore, it was also suggested that the major chromosome 9 along with other minor QTL is not responsible for more than 30% of the submergence tolerance of rice. *Sub1* was mapped to a 0.16-cM region with circa 3,000 F_2 progeny on 9th Chromosome and using 4,022 individuals a fine-scale physical mapping of *Sub1* also done, which further specified the position of locus 0.075 cM using two different types of varieties, the resistant indica Teqing and tolerant FR13A derivative IR40931-26.

Detail analysis of ERFs related to *Sub1* locus indicates the duplication of *Sub1*B leads to the development of *Sub1*A which may happen after the *indica* rice domestication. Based on variations of nucleotide in the protein coding region the *Sub1*A is divided in two allelic forms, for example, *Sub1*A-1 and *Sub1*A-2 in submergence tolerant and intolerant accessions of *indica* and *aus* lines. FR13A consists of *Sub1*A-1 which is only found in tolerant varieties, on the other hand, *Sub1*A-2 allele could be found in nontolerant *indica* accessions. Both of the allels encode identical proteins. The difference is, intolerant allele contains Pro186 and, in the tolerant allele it is Ser186. In case of expression, *Sub1*A-1 promotes rapid, prolonged, and pronounced transcript accumulation in leaves of 2- to 4-weeks-old plants in case of submerged condition; on the other hand *Sub1*A-2 promotes a lower level of transcript induction. Detailed analysis of different rice accessions with different *Sub1* haplotype showed that submergence tolerance is associated with variable levels of *Sub1*A transcript in internodes and nodes at the heading stage of the plant.

For the identification of QTL associated with waterlogging tolerance of barley different parameters was set like leaf chlorosis, plant survival rate, and biomass reduction. Then compare the QTLs identified in two seasons in two different populations using a composite map prepared by the analysis of different molecular markers like SSRs, RFLP, and DArT (Diversity Array Technology). In two barley double haploids (DH) population twenty QTLs associated with waterlogging tolerance was identified and they were validated by different replication of experiments or by different location trials. It was suggested that most of the QTLs were associated with leaf chlorosis and plant survival while some of them affected multiple waterlogging related traits. Such as QTL, Qwt4-1 contributed to overcome different waterlogging related stress like leaf chlorosis and reduction of plant biomass.

Two maize inbreed lines were crossed for the purpose of determining waterlogging tolerance, that is, "HZ32" (highly waterlogging-tolerant) and "K12" (highly waterlogging sensitive) were crossed for developing mapping population. In the maize growing season of years 2004 and 2005, two experiments were conducted, namely, EXP.1 and EXP.2, respectively, with the same families along with parents and hybrid with a treatments of controlled (without flooding) and waterlogging. One hundred seventeen SSR markers with average space of 11.5 cM were used for the preparation of population genotyping which was a base-map of 1710.5 cM length. By the use of composite interval mapping (CIM), QTL associated with different plant characters like, root dry weight, root length, plant dry weight, plant height, shoot dry weight, and waterlogging tolerance coefficient were identified in both the experiments, EXP.1 and EXP.2 in control and waterlogging condition, respectively. In EXP.1, 25 and in EXP.2, 34 QTL were identified where the effects of discovered QTLs were moderate with a range of 3.9–37.3%. Some major QTLs were identified in two chromosomes 4 and 9 and in both experiments associated with root dry weight, shoot dry weight, total dry weight, plant height, and their waterlogging tolerance coefficient. Chromosome 1, 2, 3, 6, 7, and 10 was also identified as secondary QTLs associated with tolerance.

To determine the QTL associated with waterlogging tolerance of soybean, two hundred eight lines of two recombinant inbred (RI) populations, "Archer" × "Minsoy" and "Archer" × "Noir I", were put into two different experimental setup, that is, one with control (without waterlogging) and other with waterlogged condition, and the plants of waterlogged setup were put into two weeks of flooding at early flowering stage. A single QTL was identified which was responsible for better plant growth and grain yields, 11–18% and 47–180%, respectively, in waterlogging environment linked to marker sat_064 from the parent Archer. In another experiment two populations, that is, A5403 × Archer (Population 1) and P9641 × Archer (Population 2) were used to identify QTLs associated in waterlogging tolerance where 103 and 67 RI lines, respectively, were used for mapping population. The performance of both the populations was tested on the flooded field setup and significant variations were observed, however, no transgressive tolerant segregants were found in both populations. Whole sample was divided into sensitive and tolerant bulk for the purpose of further molecular study.

In molecular genetics studies, two methods, named, SMA (single marker analysis) and CIM (composite interval mapping), were used for the identification of the QTL associated with waterlogging tolerance. Out of 912 SSR markers, 17 markers (population 1) and 15 markers (in population 2) were found associated with waterlogging tolerance in SMA. Most of the markers were closely related with *Rps* gene or QTL responsible for the resistance to Phytophthora (*Phytophthora sojae* M. J. Kaufmann and J.

W. Gerdemann). In CIM five markers, that is, Satt59, Satt160, Satt269, Satt252, and Satt485 were found associated with waterlogging tolerance in both populations. This analysis also suggested that most of the genes come from the parent Archer.

Manipulation of QTL for Developing Waterlogging Tolerant Variety

Flooding is a natural calamity that destroys production of rice throughout the world especially in low-laying areas, but all economically important varieties are not tolerant to flooding. Identification of the SUB1 QTL enables the scientist to introgress this gene by MAB into the popular high-yielding varieties. Mapping of the Sub1 allele reveals SNP (single-nucleotide polymorphisms) within *Sub1A* and *Sub1C, which* is useful for the breeders for molecular markers and in precision breeding. By the use of MAB, *Sub1A* has been already inserted to modern varieties of different countries; BR11, Swarna, Samba Mahsuri IR64, CR1009 and Thadokkam 1 (TDK1) are some of the examples of those varieties. SSR markers, which were polymorphic between two parents, were generally used for the background conformation or recurrent parent genome conformation also well combined with the Sub1 region originated from FR13A on chromosome 9. Normally newly developed Sub1 lines shows more significant waterlogging tolerance compared to parents [88, 90] which is find out by evaluation. These studies show the opportunities of the insertion of the Sub1 region from developed tolerant variety through MAB to produce tolerant varieties with a diverse genetic background. Furthermore, the effect of *Sub1* on other agronomical characteristics of plant such as, grain quality, growth, maturation, and grain production was determined in IR64 Sub1, Swarna Sub1, and Samba Mahsuri Sub1. Further research of the above mentioned submergence tolerant varieties and their original parents revealed that insertion of *Sub1* gene does not alter field performances including quality and yield of grain under normal growth condition.

Complete waterlogging of susceptible varieties at the different growing stage considerably reduces the yield attributes like, number of panicles, total number of grains per panicle, and grain filling percentage of the plant and also flowering and maturity may be delayed, resulting in a remarkable decline in yield. On the other hand, a Sub1 rice variety minimizes the total yield reduction by flood and produces more yield than the intolerant varieties at submerged condition.

One of the main advantages of using this approach is that the newly developed Sub1 varieties retain almost all agronomical characteristics of the recurrent parent, especially in terms of yield and quality. Sub1 varieties produced from FR13A-derived varieties have almost the same yield, agronomical, and grain quality characteristics as recurrent parent varieties when grown under regular condition, but, when subjected to flooding for 1 to 2 weeks, Sub1 varieties showed a remarkable advantages in terms of yield than the susceptible ones. Mega varieties which has *Sub1* gene, can be adopted by the farmer easily, in addition, these new varieties can replace the traditional landraces with low yielding, currently which has been used by the farmers in flood prone areas.

Some tolerant lines were evaluated and adopted in low-lying areas in more than ten countries in South and Southeast Asia, in field trials as the preferences of the farmers. The good yield performance of newly developed varieties, showing the better performance of *Sub1* against flooding or submergence of rice. Those performances of the newly developed varieties have influenced a lot of rice improvement programs in Asia and Southeast Asia to perform rapid seed multiplication and dissemination schemes.

Improvement of Flooding Tolerance in Rice through Plant Genetic Engineering Techniques

Current molecular biology and biotechnology methods for rice research have been improved to such level that rice has been considered as a model crop in cereal breeding. Transformation of gene of rice was firstly done by protoplast-based method through using microprojectile gun, and with the advance of technology with time *Agrobacterium* mediated gene transfer of rice is also available. The regulatory sequences of controlling expression of transgenes in rice are available now, so as a result, the transgene can be either normally expressed or expressed in response to a specific stimulus or artificially (including induced anaerobic stress, through the use of promoters from genes which strongly respond to waterlogging). Transgene rice with specific attributes like, tolerance to virus, insects. Pests, salt, low temperature, and flood can be produced. Identification/isolation/cloning of associated genes with waterlogging stress considered as limiting factor, as a result scientists focused on carbohydrate metabolism because reduced O_2 supply hampers normal respiration resulting in decrease in ATP synthesis. It has been reported that, respiration pathway switch to fermentation pathway from oxidation during oxygen-deprived (anaerobiosis) condition. The inclusion of ethanolic fermentation pathway is considered to be an important component of responses which are elicited in rice (and other plants) against flooding stress. Ethanolic fermentation, that is, pyruvate to ethanol is a relatively simpler process involving two enzymes, that is, PDC or pyruvate decarboxylase and ADH or alcohol dehydrogenase. Availability of cloned *pdc* and *adh* normally attracts the interest of molecular biologists to use these for transgenic experiments. Researchers at CISRO, Australia are trying to alter the different levels of PDC and ADH in rice and as a result 3 different *pdc* genes in rice, that is, *pdc1, pdc2,* and *pdc3* have already been cloned and sequenced. *pdc1* cDNA has been subcloned at the 3′ end of three different promoters that is, CaMV 35S, actin 1 and anoxia-induced 6X ARE promoter, which is a synthetic promoter, in both sense and antisense orientations and these plasmid constructs have been introduced into rice to produce a large number of transgenes. One of the above gene constructs (actin 1-pdc1, sense) has been employed at IRRI for the production of transgenic rice with enhanced metabolic capacity under anaerobiosis conferring submergence tolerance. In the study, tillers of transformed plants show higher *pdc* activities and ethanol production than untransformed plants and consequently ethanol production of tillers of transgenic plants was positively correlated with survival after submergence. Scientist of CISRO also made considering progress to change in the ADH level in rice.

Engineering Cold Stress Tolerance in Crop Plants

Plants respond with changes in their pattern of gene expression and protein products when exposed to low temperatures. Thus ability to adapt has an impact on the distribution and survival of the plant, and on crop yields. Many species of tropical or subtropical origin are injured or killed by non-freezing low temperatures, and exhibit various symptoms of chilling injury such as chlorosis, necrosis, or growth retardation. In contrast, chilling tolerant species are able to grow at such cold temperatures. Conventional breeding methods have met with limited success in improving the cold tolerance of important crop plants involving inter-specific or inter-generic hybridization. Recent studies involving full genome profiling/ sequencing, mutational and transgenic plant analyses, have provided a deep insight of the complex transcriptional mechanism that operates under cold stress. The alterations in expression of genes in response to cold temperatures are followed by increases in the levels of hundreds of metabolites, some of which are known to have protective effects against the damaging effects of cold stress. Various low temperature inducible genes have been isolated from plants. Most appear to be involved in tolerance to cold stress and the expression of some of them is regulated by C-repeat binding factor/ dehydration-responsive element binding (*CBF/ DREB*1) transcription factors. Numerous physiological and molecular changes occur during cold acclimation which reveals that the cold resistance is more complex than perceived and involves more than one pathway.

Abiotic stresses adversely affect growth, productivity and trigger a series of morphological, physiological, biochemical and molecular changes in plants. Cold stress is a major environmental factor that limits the agricultural productivity of plants in hilly areas. Plants respond and adapt to this stress to survive under stress conditions at the molecular and cellular levels as well as at the physiological and biochemical levels. However, expression of a variety of genes is induced by different stresses in diverse plants.

Low temperature often affects plant growth and crop productivity, which causes significant crop losses. Plants differ in their tolerance to chilling (0-15 °C) and freezing (< 0 °C) temperatures. In general, plants from temperate climatic regions are considered to be chilling tolerant with variable degree, and can increase their freezing tolerance by being exposed to chilling, non-freezing temperatures, a process known as cold acclimation, which is associated with biochemical and physiological changes and ultimately showed marked changes in gene expression, biomembrane lipid composition, and small molecule accumulation. Besides, plants of tropical and subtropical origins. are sensitive to chilling stress and lack the mechanism of cold acclimation. Low temperature resistance in plants is a very complex trait, involving many different metabolic pathways and cell compartments. Conventional breeding methods have met with limited success in improving the cold tolerance of important crop plants involving inter-specific or inter-generic hybridization. Besides, *in vitro* induced variations have also been applied to improve the abiotic stress tolerance of various crop plants but without much success. The conventional breeding approaches are limited by the complexity of stress tolerance traits, low genetic variance of yield components under stress condition and lack of efficient selection criteria. It is important, therefore, to look for alternative strategies to develop cold stress tolerant crops.

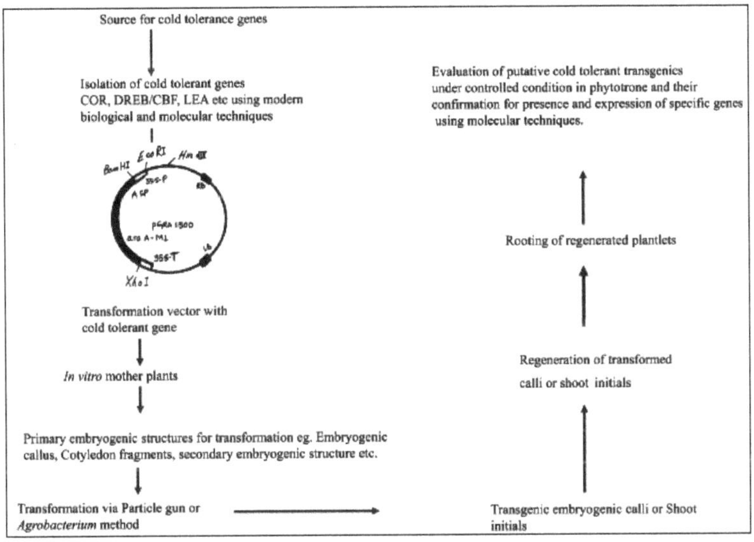

Scheme for genetic engineering of crop plants against cold stress.

Biotechnology offers new strategies that can be used to develop transgenic crop plants with improved tolerance to cold stress. Rapid advance in recombinant DNA technology and development of precise and efficient gene transfer protocols have resulted in efficient transformation and generation of transgenic lines in a number of crop species. A number of genes have been isolated and characterized that are responsive to freezing stress. Many studies have suggested that cold regulated gene expression is critical in plants for both chilling tolerance and cold acclimation. Advent of molecular tools has made it possible to select directly at the gene label without waiting for the phenotype to show up. Transgenic approach is being pursued actively throughout the world to improve traits including tolerance to biotic and abiotic stresses in a number of crops. As compared to other stresses, plant responses to cold stress are complex, so the prospects of improving cold tolerance in crops seem not to be very bright. Despite this, efforts have been made during the last two decades to generate transgenic lines of different crops, which have shown improved tolerance to cold stress. Therefore, it is important to use most appropriate tools that help in reaching the goals. The genotype designed should be better than the available ones and must reach the farmers.

Morpho-physiological basis of Cold Tolerance

A large number of studies have evaluated different plant species tolerant to different stresses such as drought, salinity and cold. However, less detail is given with regard to the methods used to evaluate the stress response; these studies may bring about some misleading conclusions from an agronomic or physiology perspective. This is particularly important, in order to closely mimic the life span of most crops under cycles of stress, rather than short exposure to very severe stresses; although we agree that short exposures to stress are certainly adequate if the purpose is to assess gene expression only. Breeding cold tolerance. Though precise details about the protocols used to evaluate the performance of plants to any given stress are very essential to assess the performance of materials.

The temperate and cool regions are those where altitudes ranged from 1600-2500 m amsl (above mean sea level) and temperature during crop growth period ranged from 5-20 °C. In temperate regions, low temperature is the primary abiotic stress which limits the crop productivity. The low temperature at seedling and reproductive stages is the major problem, results in slow establishment and low seed set which leads to poor yield of the crop. The low temperature limits the crop productivity when temperatures remain above freezing that is > 0 °C, it is called as chilling stress. Chilling sensitive cultivars are typically tropical genotypes. There is wide range of cold stress in temperate areas differing in both timing and intensity of low temperature. Yield losses are more severe when cold stress occurs during reproductive stage/ anthesis in rice which lead to high spikelet sterility. Ability of crop genotypes / lines to survive / perform better under low temperature than other genotypes is called as cold tolerance. Ordinarily, it is the consequence of cold hardening that is an earlier exposure to a low temperature for a specific period as a result of which chilling tolerance of the concerned plants increases. Cold tolerance involves increased chlorophyll accumulation, reduced sensitivity of photosynthesis, improved germination, pollen fertility and seed set which are desirable as:

Increased Chlorophyll Accumulation

Low temperature inhibits chlorophyll accumulations in actively growing leaves. In rice, cold tolerant lines, for example, *japonica* accumulates more chlorophyll under cold stress than do cold sensitive line, for example of *indica* rice. Rasolofo evaluated 181 accessions to identify donor and outstanding cold tolerant lines using leaf discolouration score (1-3) and found 19 remained green (dark) after 10 days in the 12 °C cold water tank. Sanghera found 18 cold tolerance IRCTN rice genotypes based on dark green colour and high spikelet fertility (>90%) under temperate conditions.

Reduced Sensitivity of Photosynthesis

Chloroplast and photosynthesis is major site of cold injury. Tolerance in these aspects is expressed in native vegetation adapted to growing under cool conditions. The reduced sensitivity of photosynthesis to cold has been observed in maize inbreds adapted to low temperature which is partly related to specific enzymes of the process.

Improved Germination

Genetic variation in cold tolerance at germination and seedling stage has been documented. Saini and Tandon found that L62G, Heng Jodo, Jodo, Heugdo, IRAT 102, Khonorullo, K 78-28, Daegaldo, Mujudo and L62-2A genotypes were cold tolerant having more than 85% germination and good seedling vigour (score 3) at an average of 11 °C field temperature

Improved Pollen Fertility and Seed Set

Cold tolerance at reproductive stage is expressed as improved seed set and pollen fertility. It is largely a function of floral structure and function under stress. Lia reported plant cold tolerance in rice is associated with anther size, number of pollen grain, diameter of fertile pollen grains at booting stage. However, Sanghera reported that cold tolerance is associated with high spikelet fertility (>90%) and well panicle exsertion under temperate conditions.

Cold snaps cause a reaction in the plant that prevents sugar getting to the pollen. Without sugar there is no starch build-up which provides energy for pollen germination. And without pollen, pollination cannot occur, thereby, no grain is produced. CSIRO has found that all the ingredients for starch are present but they are not getting into the pollen grain where they are needed. A cell layer surrounding the pollen, called the 'tapetum', is responsible for feeding the pollen with sugar. The tapetum is only active for 1-2 days – so if a cold snap occurs at this time, then there is no further chance for pollen growth. But the sugar cannot freely move into the tapetum and pass through it to the pollen. Instead the sugar has to be broken down then transported in bits to the pollen. 'Invertase' is the catalyst that helps in breakdown of the sugar molecule to transport it into the tapetum before it is transported to the pollen. Quantities of invertase are decreased in conventional rice when it is exposed to cold temperatures, but they remain at normal levels in a cold tolerant variety when it experiences cold. By comparing a cold tolerant strain of rice with conventional rice, CSIRO has found that the gene responsible for invertase looks exactly the same in the cold tolerant variety as it does in conventional rice. So the invertase gene itself does not make the rice plant cold tolerant – but instead a mechanism that regulates the invertase gene is different. Early research indicates that the invertase gene is regulated by the hormone abscisic acid (ABA). Oliver has experimented with injecting plants with ABA – the resulting rice plants are sterile, just like that if they experienced a cold snap. Also, ABA levels increase when conventional rice is exposed to cold, but they remain the same in the cold tolerant variety. Recent studies have indicated that the difference between cold-sensitive and tolerant rice is due to a different ability to control ABA levels. It has also been shown that this mechanism may require interactions with other plant hormones like auxins. Further, Zhao also reported that low temperature turns off the genes responsible for sugar transport into the pollen grains and therefore starch cannot be produced in the pollen in cold conditions. Cold did not cause repression of sugar delivery in cold tolerant Chinese rice and fertile pollen was still produced following cold treatment. The sugar metabolism genes also continued to function normally during cold treatment of cold tolerant rice. Ample genetic variation for cold tolerance is available in well adapted breeding population. Germplasm collected from high altitude and low temperature areas, cold tolerant mutant, somaclonal variants and wild species can be exploited for breeding improved cold tolerant genotypes in hilly areas.

Mechanisms for Understanding Tolerance to Cold Injury

Low temperature has a huge impact on the survival and geographical distribution of plants. It affects a range of cellular metabolisms in plant system depending on the intensity and duration of the stress. When exposed to low temperatures, plants respond with changes in their pattern of gene expression and protein products. Different studies have indicated that the membrane systems of the cell are the primary site of freezing injury in plants. In addition, it is well established that freeze-induced membrane damage results primarily from the severe dehydration associated with freezing. Many species of tropical or subtropical origin are injured or killed by nonfreezing low temperatures, and exhibit various symptoms of chilling injury such as chlorosis, necrosis, or growth retardation. In contrast, chilling-tolerant species are able to grow at such cold temperatures.

However, multiple forms of membrane damage can occur as a consequence of freeze induced cellular dehydration including expansion-induced-lysis, lamellar-to-hexagonal-II phase transitions, and fracture jump lesions. Thus, a key function of cold acclimation should be to stabilize membranes against freezing injury. Indeed, cold acclimation prevents expansion-induced-lyses and the formation of hexagonal II phase lipids in rye and other plants. This ability to adapt has an impact on the distribution and survival of the plant, and on crop yields. Multiple mechanisms appear to be involved in this stabilization. The best documented are changes in lipid composition. Secondly, temperature induced change in membrane fluidity is another consequences in plants during temperature stresses and might represent a potential site of perception and injury. Adaptation of living cells to chilling temperatures is a function of alteration in the membrane lipid composition by increased fatty acid unsaturation. Genetically engineered tobacco plants over-expressing chloroplast glycerol-3-phosphate acyltransferase (GPAT) gene (involved in phosphatidyl glycerol fatty acid desaturation) from squash (*Cucurbita maxima*) and *A. thaliana* showed an increase in the number of unsaturated fatty acids and a corresponding decrease in the chilling sensitivity. At low temperature, greater membrane lipid unsaturation appears to be crucial for optimum membrane function. An *Arabidopsis* fatty acid biosynthesis (*FAB1*) mutant with more saturated membranes showed decreased quantum efficiency of photosystem II (PSII), chlorophyll content and the amount of chloroplast glycerolipids after prolonged exposure to low temperature. A triple mutant fatty acid desaturation (*fad3-2 fad7-2 fad8*) devoid of trienoic fatty acids (18:3 or 16:3) produced a phenotype similar to *FAB1*, when plants were subjected to prolonged low temperature exposure.

Similarly, *fad5*and *fad6* mutants with more saturated membranes became chlorotic and showed growth retardation during low temperature incubation. In addition to membrane unsaturation, it appears that lipid asymmetry in the membrane also contributes to membrane physical structure at low temperature. The accumulation of sucrose and other simple sugars that typically occurs with cold acclimation also seems likely to contribute to the stabilization of membranes as these molecules can protect membranes against freeze-induced damage *in vitro*.

Besides, protective chaperone like function of LEA proteins acting against cellular damage has been proposed, indicating the role of LEA proteins in anti aggregation of enzymes under desiccation and freezing stresses. There is emerging evidence that certain novel hydrophilic and late embryogenesis abundant (LEA) polypeptides also participate in the stabilization of membranes against freeze-induced injury. These hydrophilic and late embryogenesis abundant polypeptides are predicted to contain regions capable of forming amphipathic α-helices which are shown to have strong effect on intrinsic curvature of monolayers and their propensity to form hexagonal II phase. They are said to defer their formation at lower temperatures. There is another evidence that freeze-induced production of reactive oxygen species contributes to membrane damage and that intercellular ice can form adhesions with cell walls and membranes and cause cell rupture. Further, there is evidence that protein denaturation occurs in plants at low temperature which could potentially result in cellular damage. In these cases, the enhancement of antioxidative mechanisms, increased levels of sugars in the apoplastic space, and the induction of genes encoding molecular chaperones, respectively, could have protective effects.

In cold stress-tolerant plants, many genes involved in the synthesis of osmoprotectants—organic compounds such as amino acids (e.g. proline), quaternary and other amines (e.g. glycinebetaine and polyamines) and a variety of sugars and sugar alcohols (e.g. mannitol, trehalose and galactinol) that accumulate during osmotic adjustment—have been used. Both cold-stress-induced transcripts and constitutively expressed transcripts need to be processed, exported to the cytoplasm and kept in conformations that are competent for translation. RNA can fold into extensive secondary structures that could interfere with its function, and cold temperatures exacerbate this interference. Many genes that respond to multiple stresses like dehydration and low temperature at the transcriptional level are also induced which protects the cell from dehydration and chilling. In order to restore the cellular function and make plants more tolerant to stress, transferring a single gene encoding a single specific stress protein may not be sufficient to reach the required tolerance levels. To overcome such constraints, enhancing tolerance towards multiple stresses by a gene encoding a stress inducible transcription factor that regulates a number of other genes is a promising approach. In bacteria, nucleic-acid-binding cold shock proteins (CSPs) accumulate at cold temperatures and function as transcription antiterminators or translational enhancers by destabilizing RNA secondary structure. Some CSP-domain-containing proteins in plants are upregulated by cold stress, and might function as RNA chaperones in the regulation of translation. A different cold-responsive nucleic-acid-binding protein, a zincfinger-containing glycine-rich RNA-binding protein from *Arabidopsis* designated *atRZ-1a*, is also upregulated by cold stress, and genetic analysis supports its function in freezing tolerance. Compared to other organisms, plants have the largest number of DEAD-box RNA helicase genes. One of these helicases, which is encoded by the *Arabidopsis* low expression of osmotically responsive genes4 (*LOS4*) gene, is essential for plant tolerance of chilling and freezing stress. *LOS4* is required for efficient export of RNA from the nucleus to the cytoplasm. The *Arabidopsis* nucleoporin *AtNUP160* suppressor of auxin resistance1 (*SAR1*) also controls RNA export, and is crucial for chilling and freezing tolerance. Both *LOS4* and *AtNUP160*proteins are enriched at the nuclear rim. Defects in the nucleocytoplasmic transport of RNA seem to affect cold tolerance preferentially, because the *LOS4* and *AtNUP160* mutant plants do not have severe growth or developmental phenotypes, nor are they strongly altered in the tolerance of other abiotic stresses.

Key Players Involved in Cold Responsive Pathways

Cold tolerance is the result of complex physiological mechanisms involving many cell and plant traits. Earlier studies have shown that the genetic control of cold tolerance is complex and can be regarded as polygenic and the mechanism of how these genes controlled cold tolerance is still not fully clear. Therefore, the foundation for a better molecular and genetic understanding of the cold responsive pathways will improve our knowledge and pave the way for the development of improved methodologies for cold tolerance screening. Key to the tolerance of plants to abiotic stresses is a complex network of transcription factors and other regulatory genes that control multiple defense enzymes, proteins and pathways. The discovery of gene expression change during cold acclimation was the starting of exploration of antifreezing molecular mechanisms. In this context, Zhao reported that gene expression profiling using DNA chips indicated that large numbers of genes were differentially expressed under cold stress. They identified 242 unique genes that are expressed differentially between cold sensitive and cold tolerant rice. These genes are involved in processes such as senescence, cell death, male sterility and plant hormone response. Similarly, global transcript profiling analyses indicated that > 10% of genes in the *Arabidopsis* genome were regulated during cold acclimation. Transcriptome analysis using microarray technology is a powerful technique, which has proven very useful for discovering many stress-inducible genes involved in stress response and tolerance. Genes involved in stress signal sensing and a cascade of stress-signaling in *A. thaliana* has been of recent research interest. Components of the same signal transduction pathway may also be shared by various stress factors such as drought, salt and cold. Although there are multiple pathways of signal-transduction systems operating at the cellular level for gene regulation.

In past, it has been reported that genes induced during stress conditions function not only in protecting cells from stress by producing important metabolic proteins, but also in regulating genes for signal transduction in the stress response. Several stress induced cor genes such as rd29A, cor15A, kin1 and cor6.6 are triggered in response to cold treatment, ABA and water deficit stress in the early stages of the osmotic stress response.

Several stress induced cor genes such as rd29A, cor15A, kin1 and cor6.6 are triggered in response to cold treatment, ABA and water deficit stress in the early stages of the osmotic stress response. Similarly, a cis-acting element, dehydration responsive element (DRE) identified in A. thaliana, is also involved in ABA-independent gene expression under drought, low temperature and high salt stress conditions in many dehydration responsive genes like rd29A that are responsible for dehydration and cold-induced gene expression. Thus, clearly, the overexpression of some drought-responsive transcription factors can lead to the expression of downstream genes and the enhancement of abiotic stress tolerance in plants. Thus, these gene products are classified into two groups.

The first group of proteins that probably function in stress tolerance includes chaperones, LEA proteins, osmotin, antifreeze proteins, mRNA-binding proteins, some key enzymes for osmolyte biosynthesis (like proline, water channel proteins, sugar and proline transporters, detoxification enzymes), enzymes for fatty acid metabolism (proteinase inhibitors, ferritin) and lipid-transfer proteins. It has been reported that some of these stress-inducible genes specially encoding proteins (such as enzymes for osmolyte biosynthesis, LEA proteins and detoxification enzymes) have been overexpressed in transgenic plants and produce stress-tolerant phenotypes in the transgenic plants. These results indicate that the gene products of the stress-inducible genes really function in stress tolerance. The second group contains protein factors involved in regulation of signal transduction and gene expression that probably function during stress response. This group includes various transcription factors that regulate different stress-inducible genes collectively or separately, and may constitute gene networks. Seki reported that some of these regulatory pathways are also involved in drought-, cold-, or high-salinity stress responses. Though, the clear cut functions of most of these genes are not fully understood. Functional analysis of these stress-inducible transcription factors will provide precise information on the complex regulatory gene networks that are involved in responses to drought, cold, and high-salinity stresses. Some of these stress-inducible regulatory genes that encode proteins (transcription factors) have been overexpressed in transgenic plants and generate stress-tolerant phenotypes in them.

Cold Tolerance using Transgenic Approaches

When a plant is subjected to abiotic stress, a number of genes are turned on, resulting in increased levels of several metabolites and proteins, some of which may be responsible for conferring a certain degree of protection to these stresses. A key to progress towards breeding better crops under stress has been to understand the changes in cellular, biochemical and molecular machinery that occur in response to stress. A key to progress towards breeding better crops under stress has been to understand the changes in cellular, biochemical and molecular machinery that occur in response to stress which in turn provides new tools and strategies to improve the environmental stress tolerance of crops. Since freezing tolerance is a multigenic trait, transformation of a single functional gene appears to have a limited effect on crop freezing tolerance. Because many aspects of cold adaptation process are under transcriptional control, many transcription regulatory factors were chosen, hence, genetic engineering for introgression of such genes that are known to be involved in stress response and putative tolerance, might prove to be a faster track towards improving crop varieties for enhanced cold tolerance.

Low-temperature limitations have been overcome by the identification of cold-tolerant genes for applications in genetically transformed crops. In transgenic tobacco (*Nicotiana tabacum*), chilling tolerance at 1 °C for 7 d was achieved by the over expression of a gene encoding chloroplast *w*3 fatty acid desaturase. Furthermore, tolerance at 1 °C for 11 d was conferred using a gene encoding a non-specific cyanobacterial desaturase, and the resultant transgenic tobacco plants showed a reduction in saturated fatty acid content in membrane lipids. The over expression of glycerol-3-phosphate acyl transferase altered the unsaturation of fatty acids and conferred chilling tolerance in transgenic plants. Hence, modifications in lipid composition that stabilize cell membranes and prevent cellular leakage lead to cold tolerance.

Transgenic technology has opened up many exciting possibilities to improve cold stress in plants by introduction or removal of gene or genes that regulate a specific trait. It also offers uncommon opportunities for improvement in genetic potential of plants in the form of development of specific crop varieties that are more resistant to biotic and cold stresses with enhanced nutritional level.

During last two decades, advancement in plant biotechnology has led to the identification and isolation of a number of transcription factors related to cold stress tolerance. A good number of genes which have been identified in different studies raise the question of exactly which genes are most central to increasing cold/ freezing tolerance. The genes selected for transformation should be involved in encoding enzymes that are required for the biosynthesis of various osmoprotectants. Other classes of genes that selected for transformation include those that encoded enzymes for modifying membrane lipids, LEA protein, and detoxification enzymes. In these studies, either a single gene for a protective protein or an enzyme was overexpressed under the control of the constitutive 35S cauliflower mosaic virus (*CaMV*) promoter in transgenic plants, although several genes have been shown to function in environmental stress tolerance and response. The genes encoding protein factors that regulate gene expression and signal transduction, that function in stress responses may be useful for improving the cold tolerance of plants by gene transfer as they can regulate many stress-inducible genes involved in cold stress tolerance.

Selective Reports on Production of Cold Stress-tolerant Transgenic Crops

The CBF genes represent one of the most significant discoveries in the field of low temperature adaptation and signal transduction. All important crops and few vegetables species have contained this gene. Various low temperature-inducible genes have been isolated from plants that appear to be involved in tolerance to cold stress, and the expression of some of them is regulated by C-repeat binding factor/dehydration-responsive element binding (*CBF/ DREB1*) transcription factors. Three *CBF/DREB1* genes (*CBF3/DREB1a, CBF1/DREB1b*, and *CBF2/DREB1c*) belonging to the *AP2/DREBP* family of DNA-binding proteins have been identified in *Arabidopsis*. Transgenic *Arabidopsis* plants constitutively over-expressing a cold inducible transcription factor (*CBF1; CRT/DRE* binding protein) showed tolerance to freezing without any negative effect on the development and growth characteristics. Overexpression of *Arabidopsis CBF1* (*CRT/DRE* binding protein) has been shown to activate cor homologous genes at non-acclimating temperatures. The *CBF1* cDNA when introduced into tomato (*Solanum lycopersicum*) under the control of a CaMV35S promoter improved tolerance to chilling, drought and salt stress but exhibited dwarf phenotype and reduction in fruit set and seed number per fruit.

The expression of related cold shock proteins (CSPs) from bacteria, CspA from *Escherichia coli* and CspB from *Bacillus subtilis*, promotes stress adaptation in multiple plant species. Transgenic rice plants expressing CspA and CspB manifest improved stress tolerance for a number of abiotic stresses, including cold, heat, and water deficits. It has long been established that changes in gene expression occur upon exposure to cold acclimation. Number of COR genes isolated from *Arabidopsis* that encode polypeptides thought to have protective roles against dehydration. Expression profile experiments in *Arabidopsis* demonstrated that extensive changes in gene expression occur during cold acclimation and that a substantial number of the genes that are up-regulated by the cold response are involved in metabolism. Hajela reported that a set of genes that encode a related family of cold-regulated (*COR*) proteins, which are massively induced during cold acclimation. These *COR* genes were used by different researchers to identify a family of *Arabidopsis* transcription factors known as either C-repeat binding factors (*CBF*) (*CBF1, CBF2 and CBF3*) or dehydration responsive element-binding factors (*DREB*) (*DREB1B, DREB1C* and *DREB1A*). In Arabidopsis (*A. thaliana*) and rice, the CBF/DREB1-dependent cold response pathway has been shown to play a predominant role in freezing tolerance through the process of cold acclimation. These *CBFs/DREBs* are upstream transcription factors that bind to promotor *cis*-element *CRT/DRE* and activate the expression of these cold responsive genes. Overexpression of *CBF1/DREB1b* and *CBF3/DREB1a* enhances cold tolerance by inducing *COR* (cold regulated) genes. Furthermore, its overexpression causes many biochemical changes, such as the accumulation of sugar and proline. Thus, the *CBF/DREB1* genes are thought to be activators that integrate several components of the cold acclimation response by which plants increase their tolerance to low temperatures after exposure to nonfreezing conditions. Several stress induced *COR* genes such as *rd*29A, *COR*15A, *kin*1 and *COR*6.6 are triggered in response to cold treatment, ABA and water deficit stress.

There have been numerous efforts in enhancing tolerance towards multiple stresses such as cold, drought and salt stress in crops other than the model plants like *Arabidopsis*, tobacco and alfalfa. An increased tolerance to freezing and drought in *Arabidopsis* was achieved by overexpressing *CBF*4, a close *CBF/ DREB*1 homolog whose expression is rapidly induced during drought stress and by ABA treatment, but not by cold. Subsequently, the overexpression of *DREB*1A has been shown to improve the drought- and low-temperature stress tolerance in tobacco, wheat and groundnut. The use of stress inducible *rd*29A promoter minimized the negative effects on plant growth in these crop species. Further, Gilmour have shown that ectopic transgenic overexpression of *CBF1/ DREB1B*, *CBF2/ DREB1C* or *CBF3/DREB1A* in *Arabidopsis* activates a suite of *CBF/ DREB* target genes at warm temperatures and results in increased freezing and drought tolerance. *CBF* transcripts begin accumulating within 15 minutes of plants being exposed to low temperature strongly suggests that the low temperature "Thermometer" and "Signal Transducer" are present at warm non-inducing temperatures. Two cDNA clones that encode DRE-binding proteins, *DREB1A/CBF3* and *DREB2A*, have been isolated by using yeast one-hybrid screening. cDNA clones encoding two *DREB1A* homologs (named *DREB1B/CBF1* and *DREB1C/CBF2*) and one *DREB2A* homolog (*DREB2B*) were also isolated and Expression of the *DREB1A* gene and its two homologs was induced by low-temperature stress, whereas expression of the *DREB2A* gene and its homolog was induced by dehydration. These results indicate that two independent families of DREB proteins, *DREB1* and *DREB2*, function as transacting factors in two separate signal transduction pathways under low-temperature and dehydration conditions, respectively.

*CBF*s are negatively regulated by an upstream transcription factor, *MYB15* (an *R2R3-MYB* family protein) in *Arabidopsis*. *MYB15* is expressed even in the absence of cold stress, and *MYB15* can bind to *MYB* recognition elements (*MYBRS*) in the promoters of *CBFs*. *MYB15* mutant plants show enhanced expression of CBFs during cold acclimation and enhanced freezing tolerance, whereas, transgenic *Arabidopsis* overexpressing *MYB15* showed a decreased expression of *CBFs*and a reduction in freezing tolerance. Thus, *MYB15* is an upstream transcription factor that negatively regulates the expression of *CBFs*.

Transcriptome analysis of *ZAT12*-overexpressing *Arabidopsis* revealed that the *ZAT12* regulon consists of at least 24 cold standard set (*COS*) genes, of which nine are cold-induced and 15 are cold-repressed genes. Constitutive overexpression of *Arabidopsis DREB1A* improved drought and low-temperature stress tolerance in tobacco, and regulation of transgene expression *via* the stress-inducible *RD29A* promoter minimized the negative effects on plant growth. Similarly, the *Arabidopsis DREB1A* gene was placed under control of the *RD29A* promoter and transferred *via* biolistic transformation into bread wheat. However, constitutive overexpression of the *CBF* genes using the cauliflower mosaic virus 35S promoter can result in undesirable agronomic traits. In *Arabidopsis*, *CBF* overexpression can cause a "stunted" growth phenotype, a decrease in seed yield and a delay in flowering.

In the last decade, extensive research efforts have been undertaken to identify and characterize cold-responsive (*COR*) genes and a number of homologous components of the *Arabidopsis CBF* cold response pathway in many plants have been found. Many of these putative orthologs have been structured, analyzed and functionally tested. The expression patterns of the *CBFs* and *CORs* in response to low temperature are similar in a variety of plants species, involving rapid cold-induced expression of the *CBF*s followed by expression of *CBF*-targeted genes that increase freezing tolerance. Moreover, constitutive overexpression of the *Arabidopsis CBF* genes in other plants resulted in increased freezing tolerance that have been successfully used to engineer cold stress tolerance in several crop species.

Transgenic attempts with other structural genes have also been made with fair degree of success. Genetically engineered tobacco plants over-expressing chloroplast glycerol-3-phosphate acyltransferase (*GPAT*) gene (involved in phosphatidyl glycerol fatty acid desaturation) from squash (*Cucurbita maxima*) and *A. thaliana* showed an increase in the number of unsaturated fatty acids present in the plant cell wall, which enhance the cold tolerance to the plants during cold stress. Expression of a plant phosphatase (*At PP2CA*) in transgenic *A. thaliana* can accelerate the development of cold acclimation and increase freezing tolerance. It has also shown that transgenic plants expressing a constitutively active kinase *NPK1*, is more tolerant to chilling and other abiotic stresses. Pennycooke reported that down-regulating α-Gal (α -Galactosidase) in transgenic petunia resulted in an increase in freezing tolerance suggesting that engineering raffinose metabolism by transformation with α -Gal provides an additional method for improving the freezing tolerance of plants. The overexpression of genes encoding *LEA* proteins can improve the stress tolerance of transgenic plants. Expression of the citrus gene encoding a *LEA* protein, *CuCOR19* increased the cold tolerance of transgenic tobacco. Likewise, the freezing tolerance of *Arabidopsis* was increased by the ectopic expression of the wheat gene *WCS19*, the *Arabidopsis* gene *COR15A*, and the co-expression of the genes *RAB18* and *COR47* and *XERO2* and *ERD10*. The freezing tolerance of strawberry leaves was enhanced by expression of the wheat dehydrin gene *WCOR410*. On the other hand, the expression of two cold-induced *LEA* proteins from spinach and three desiccation-induced *LEA* proteins from *C. plantagineum* in tobacco did not induce any significant changes in the freezing or drought tolerance of the respective transgenic plants. This may indicate either that not all *LEA* proteins make a significant contribution to plant stress tolerance, or that they need a particular background to function in, as suggested for transgenic strawberry plants. Kim engineered tobacco with ring zinc finger protein (*RDCPt*) from hot pepper (*Capsicum annuum*) and reported that expression of this gene resulted in improved cold tolerance in transgenic plants as compared to wild type. Dai reported that overexpression of

OsMYB3R-2 in transgenic *Arabidopsis* increased tolerance to freezing when exposed to -8 °C for 10 h. They found that survival after 6 d at normal conditions was 26.8% for the wild-type and 84.5% for transgenic lines. Phenotypically, most transgenic seedlings were green and could regrow as compared with the wild type; whereas most wild-type seedlings became white and did not regrow after removed to normal conditions. The survival percentage under different low temperatures also showed dramatic difference between the transgenic plants and wild-type plants.

Pramanik and Imai reported that *TPP* (trehalose-6-phosphate phosphatase) genes expressed in rice and their expression is induced by cold. Trehalose accumulates rapidly and transiently, which follows the transient induction of *TPP* activity, in rice tissues during chilling stress. Overexpression of *TPS(trehalose-6-phosphate synthase)* and *TPP* genes enhanced the accumulation of trehalose and tolerance to cold stress in transgenic tobacco and rice. However, the regulatory mechanism of *TPPs* by cold or other stresses is unclear. *MYBS3* plays a critical role in cold adaptation in rice and necessary for enhancing cold tolerance. They reported that transgenic rice constitutively overexpressing *MYBS3* tolerated 4 °C for at least 1 week and exhibited no yield penalty in normal field conditions.

Based on previous studies it has been established that the *CBF* cold responsive pathway is an integral component of the cold acclimation response. However, the transcriptome data showed that additional cold-regulatory pathways also exist. Transcriptome comparisons indicated that only 12% of the cold-responsive genes are certain members of the *CBF* regulon. Moreover, at least 28% of the cold-responsive genes were not regulated by the *CBF* transcription factors, including 15 encoding known or putative transcription factors, indicating that these cold-responsive genes are members of different low-temperature regulons.

When overexpressed in *Arabidopsis* and tobacco, the soybean gene *SCOF-1* (encodes a zinc-finger protein) can activate *COR* gene expression and increase freezing tolerance in non-acclimated transgenic plants, although the *SCOF-1* protein does not directly bind to either the *DRE/CRT* or the ABRE elements. *SCOF-1* interacts with another G-box binding *bZIP* protein, *SGBF-1*. *SGBF-1* can activate ABRE-driven reporter gene expression in *Arabidopsis* leaf protoplasts. Thus, *SCOF-1* may regulate the activity of SGBF-1 as a transcription factor in inducing *COR* gene expression.

Forward genetic analysis in *Arabidopsis* identified two transcription factors, high expression of osmotically responsive genes, *HOS9* and *HOS10*, which are required for basal freezing tolerance. Similarly, microarray analysis led to the identification of the cold-stress-inducible *AP2* family transcription factor gene related to *ABI3/VP1* (*RAV1*) that might regulate plant growth under cold stress. *RAV1* is down regulated by epibrassinolide and transgenic *Arabidopsis* overexpressing *RAV1* exhibits a retardation of lateral root and rosette-leaf development. Further, the importance of *CBF*-independent pathways is also supported by analysis of mutants that have increased freezing tolerance, for example, mutations in eskimo1 (*ESK1*), a protein of unknown function, result in constitutive freezing tolerance. The *Eskimo1* mutation was first identified as a mutation conferring frost survival without an acclimation period. Later on, Bouchabke-Coussa clearly showed that *ESKIMO1* mutants are more tolerant to freezing but only after acclimation. The genes that are affected by the *ESK1* mutation are distinct from those of the *CBF*regulon. Recently, a *CBF* 1 gene has been introduced into tomato that resulted in transgenic plants showing higher activity of superoxide dismutase (SOD), higher non-photochemical quenching (NPQ), and lower malondialdehyde (MDA) content. Thus suggesting that *CBF1* protein plays an important role in protection of PSII and PSI during low temperature stress at low irradiance.

Aforementioned studies clearly demonstrate the complex and interactive relationships among different pathways regulated by cold acclimation. Hence, elucidation of the mechanism regulating cold-regulated genes is an important goal in achieving a full understanding of cold acclimation. With the advent of molecular genetics and biotechnology, it is now possible to genetically engineer plants to be more tolerant to low temperature. The technologies can be so developed/ modified that will significantly improve breeding efficiency, resulting in rapid and accurate incorporation of cold tolerant genes into crop plants.

Prospects of Cold Tolerant Genes in Hill Agriculture

Cold is an environmental factor that limits the geographical distribution and growing season of many plant species, and it often adversely affects crop quality and productivity. Most temperate plants can acquire tolerance to freezing temperature by a prior exposure to low non-freezing temperature, a process known as cold acclimatization. Plants of tropical and subtropical origins are sensitive to chilling temperature (0°C-10°C) and are incapable of cold acclimation. Genes induced by stress can be roughly classified into two groups: genes coding for regulatory proteins, mainly transcription factors, and genes encoding proteins involved directly in response mechanisms; genes from both classes are of interest. Variations in the expression of regulators could lead to a protective status before the emergence of stress and have multiple effects. Genes involved in protection or repair mechanisms could be new targets for the improvement of plant plasticity and adaptive responses to stress. The unraveling of general stress responses in the model species *Arabidopsis thaliana* helped to identify potential targets for plant breeding. *Arabidopsis* genes involved in tolerance to abiotic stress were transferred, by genetic engineering, to many crops and tolerance was successfully conferred in the field, despite the complexity of plant responses to environmental stress. Thus, finding new key genes responsible for abiotic stress tolerance phenotypes is of great importance not only for a better understanding of stress responses, but also for promising future crop improvement.

Many studies have suggested that cold regulated gene expression is critical in plants for both chilling tolerance and cold acclimation. Cold responsive genes encode a diverse array of proteins such as enzymes involved in respiration and metabolism of carbohydrates, lipids, phenylpropanoids and antioxidants: molecular chaperones, antifreeze proteins, and others with a presumed function in tolerance to dehydration caused by freezing). The change in the gene expression occur in plant during cold acclimatization a developmental process that results in increased tolerance. Since then, it has repeatedly been speculated that certain *COR* (cold regulated) genes might have role in freezing tolerance. To test this notion investigators have turned to isolating the characterizing genes that are expressed in response to low temperature. These efforts have led to the identification of a number of genes such as the *COR 15a KINI, LTI 78, fad 7*etc. of *A. thaliana*. The generic trends of the genes and transcription factors are available in STIFDB (Stress responsive Transcription Factor Database). STIFDB is a database of stress-related genes, which are upregulated in abiotic stress-related microarray experiments. STIFDB provides a platform to understand the stress-regulome of abiotic stress responsive genes in plants. STIFDB will be a highly useful resource for a researcher working on abiotic stress responses in plants.

Low temperature stress is a major environmental factor that not only limits where crops can be grown but also reduces yields depending on the weather in a particular growing season. In addition to exceptionally stressful years that cause significant yield reductions, less extreme stress almost certainly causes smaller losses over large areas to produce comparable yield reductions every year. Even in cases when freezing stress does not result in yield loses; it often results in crop quality reduction. Each year, worldwide losses in crop production due to low temperature damage amount to approximately $ 2 Billion. Some of the major losses include the 1995 early fall frosts in the US which caused losses of over $1 billion to corn and soybeans. The occasional freezes in Florida have shifted the citrus belt further south, and California sustained $650M of damage in 1998 to the citrus crop due to a winter freeze. The inherent cold hardiness of the crop determines in which agricultural areas it can be grown. Crops that are more resistant to freezing stress would allow some geographical regions to grow more profitable and productive crops with less environmental risks. However despite continued efforts, traditional breeding has had only limited success in imparting crop plants with better freezing tolerance due to very little was known about the mechanisms that regulate chilling and freezing tolerance. With the advent of molecular genetics and biotechnology, it is now possible to genetically engineer plants to be more tolerant to many environmental adversities, including low temperature. Molecular studies have shown that several genes with various functions are induced by environmental stresses such as drought, high-salinity and low temperature in plants. Most of the dehydration responsive genes are induced by the plant hormone abscisic acid (ABA), but others are not. Expression analyses of dehydration-responsive genes have provided at least four independent regulatory systems (regulons) for gene expression in a model plant *Arabidopsis thaliana*. The *cis*-acting elements in the promoters of some genes that have a typical stress-inducible expression profile and the transcription factors that affect the expression of these genes have been analysed.

Breeding for Salinity Tolerance in Plants

Plants are subjected to various abiotic stresses such as low temperature, salt, drought, floods, heat, oxidative stress and heavy metal toxicity during their life cycle. Among all this, salinity is the most typical abiotic stress. Salinity has negative impact on agricultural productivity affecting plant growth and restricting the use of land. It is estimated that 6% of the world's total land and 20% of the world's irrigated areas are affected by salinity. World population is increasing at an alarming rate and is supposed to reach nine billion by 2050, but our food production is limited. As green revolution has already reached its ceiling, there is a major concern over food supply for the ever increasing world population. Rapidly shrinking agricultural land, due to industrialization and habitat use is a major threat to sustainable food production. In light of all this, it is almost imperative to raise salt tolerant plants to effectively use salt affected agricultural land for sustainable crop production. Salinity is a soil condition characterized by a high concentration of soluble salts. Soils are classified as saline when ion s concentration is such that osmotic pressure produced by ion s are equivalent to that generated by 40 mM NaCl i.e. 0.2 MPa or more. As NaCl is the most soluble and widespread salt, it is not surprising that all plants have evolved mechanisms to regulate its accumulation and to select against it in favor of other nutrients commonly present in low concentrations, such as K^+ and NO_3^-. Salinity problem is further aggravated by irrigation and is more in hot temperate regions, where there is excessive water loss through transpiration. The initial effect of salt stress is osmotic stress caused by the presence of ions in rhizosphere which restricts extraction of water by roots and results in reduced plant growth.

The secondary effects of salt stress are caused by ionic disequilibrium, which result in inactivation of enzymes, nutrient starvation, ionic toxicity in tissues and oxidative stress. Reactive oxygen species produced due to oxidative stress further damage plants by enhancing lipid peroxidation, DNA damage and inhibition of photosynthesis. If the concentration of salt is very high, it leads to plant death. There is inter-species and intra-species variability in salinity tolerance in plants. Plants adopt different mechanism to resist salinity stress like excluding salts or accumulating ions into different tissue compartments, vacuoles or old leaves. In most plants, Na^+ and Cl^- are effectively excluded by roots, while water is taken up from the soil. In response to osmotic stress, plants produce osmolytes like glycine betaine, trehalose or proline, which protect them from dehydration or protein denaturation. However, oxidative stress-an outcome of ionic stress lead to the production of different enzymatic or non-enzymatic antioxidants, which protect plants from harmful effects of reactive oxygen species. Plant breeding is being used since long to produce salt tolerant and more productive lines. However, its use is limited due to multigenic nature of salt tolerance and presence of low genetic variation in major crops. In recent times, genetic engineering has played a pivotal role in producing salt tolerant plants.

Conventional Breeding Approach

Plant breeding has been used since long for the production of high yielding and stress tolerant crops. Plant breeders have used genetic variation in crops, at intraspecific, interspecific and intergeneric levels to produce salt tolerant lines. However, breeding has the limitation due to low magnitude of variation in gene pools of most crops. Another problem associated with conventional breeding, is that if the gene is present in a wild relative of the crop, there is difficulty in transferring it to the domesticated cultivar, due to reproductive barrier.

Role of QTL and Molecular Markers in Engineering Salinity Tolerance

QTLs are segment of genetic material, in the genome of an organism linked with a particular trait. Salt stress tolerance is a complex trait, so the QTLs related with salt tolerance have significant role in understanding the stress response and generating stress-tolerant plants. There has been progress in methods of identifying genes underlying QTLs, instead of only map based cloning approach; new approaches like microarray based transcriptional profiling of differential gene expression or combination of genetic mapping and expression profiling are being used for identifying genes linked with QTLs. Several QTLs involved in the salt stress responses, have been reported of recent. The recent developments in molecular marker analysis have made it feasible to analyze both simply inherited; as well as the quantitative traits, and identify individual gene controlling the trait of interest. Molecular markers could be used to tag quantitative trait loci and to evaluate their contributions to the phenotype by selecting for favorable alleles at these loci in a marker-aided selection scheme aiming to accelerate genetic advance. Advanced backcross QTL analysis can be used to evaluate mapped donor introgression in the genetic background of an elite recurrent parent.

Genetic Engineering Approach for Salinity Stress Tolerance

Plant breeding strategy for salt tolerance is not much successful due to reproductive barrier and also as it involves the risk of other undesirable traits transfer. So to avoid this problem, genetic engineering strategy is more preferred, as it only deals with the specific gene s transferred. Plants try to cope with salinity by inducing various metabolic changes like production of osmolytes, antioxidative enzymes and upregulating various genes involved in stress response like ion transporters, ion channels, transcriptional factors and various signaling pathway components. These plant responses to salinity have been utilized by scientists to generate transgenic; either by, transferring such stress responsive gene s to salt-sensitive crop plant from different genetic background (relatively salt-tolerant plants) or altering the expression of existing genes.

There are a number of gene s known which are responsible for salinity tolerance when transferred in plants through genetic engineering.

Genes for Osmoprotectants

When plants are exposed to stress conditions, metabolic shifts occur and result in changes in the levels of a various cellular metabolites. Such modifications in response to abiotic stress appear to be associated with the enhanced ability to tolerate such stressful conditions. Metabolites that might be expected to contribute to enhanced salt stress tolerance include soluble sugars, amino acids, organic acids, polyamines and lipids. One important group of such metabolites is so-called 'compatible solutes', which are small organic metabolites that are very soluble in water and are non-toxic at high concentrations.

Trehalose

Trehalose, a nonreducing disaccharide plays a crucial role in metabolic homeostasis and abiotic stress tolerance in various organisms. In plants, trehalose-6-phosphate synthase (TPS) catalyzes the transfer of glucose from UDP-glucose to glucose-6-phosphate (G-6-P) to form trehalose-6-phosphate (T-6-P) and uridine diphosphate (UDP). Subsequently, the T6-P is dephosphorylated into trehalose by trehalose-6- phosphate phosphatase (TPP). Li , have shown that overexpression of OsTPS1 gene encoding trehalose-6- phosphate synthase in rice improved the tolerance of rice to high salinity and other abiotic stresses. Overexpression of this gene in rice is associated with increased level of trehalose and proline along with upregulation of some of the stress inducible genes including WSI18, RAB16C, HSP70 and ELIP. Transfer of the yeast TPS1 into tomato resulted in higher chlorophyll, starch content and enhanced tolerance against drought, salt and oxidative stresses. Rice plants transformed with Escherichia coli's trehalose biosynthetic gene s (otsA and otsB) as a fusion gene exhibits less photo-oxidative damage and a more favorable mineral balance under salt, drought and low-temperature stress conditions. Similarly, in tobacco, heterologous expression of AtTPS1 gene from Arabidopsis increased tolerance to several abiotic stresses such as drought, desiccation and temperature stresses. However, the gene transfer for trehalose can also produce aberrations in plant growth such as dwarfism, delayed flowering, abnormal root development and lancet-shaped leaves.

Glycine Betaine

Glycine betaine (N, N, N-trimethyl glycine) is a quaternary ammonium compound found in bacteria, haemophilic archaebacteria, marine invertebrates, plants and mammals. GB is synthesized; either by, the oxidation (or dehydrogenation) of choline or by the N-methylation of glycine. It accumulates to osmotically significant levels in many salttolerant plants and halotolerant cyanobacteria. Levels of GB vary considerably among plant species and organs. Plants of many taxonomically distant species normally, contain low levels of GB (these plants are known as natural accumulators of GB), but they accumulate larger amounts of GB when subjected to abiotic stress. In many other species GB is not detectable under normal or stressful conditions. There are now strong evidences that GB plays an important role in abiotic stress tolerance. The biological functions of GB have been studied extensively in higher plants such as spinach, sugar beet, barley and maize. The availability of GBaccumulating transgenic plants has provided insight into its plant cell protection mechanism. Furthermore, many lines of GB-accumulating transgenic plants exhibit greatly improved tolerance to various types of abiotic stresses and their properties suggest promising strategies for the development of stress-tolerant crop plants. Gene s that encoding GBbiosynthetic enzymes have been cloned from different organisms to generate transgenic plants. The transgenic plants accumulate GB at different levels and exhibit enhanced tolerance to salt and other abiotic stresses. Exogenous application of glycine betaine improves salinity tolerance in many plant species enhancing plant growth and yield. Transgenic tomato and rice expressing codA gene from Arthrobacter gobiformis show enhanced salinity tolerance. Similarly,; transgenic rice plants, for cox gene coding for choline oxidase from Arthrobacter pascens were found salt tolerant. Genetically engineered tobacco (Nicotiana tabacum) plants for betA gene from E.coli coding for choline dehydrogenase exhibit salt tolerance. In the same vein, transgenic plants for enhanced synthesis of glycine betaine have also been produced in Brassica, Arabidopsis and Solanum tuberosum showing enhanced salinity tolerance.

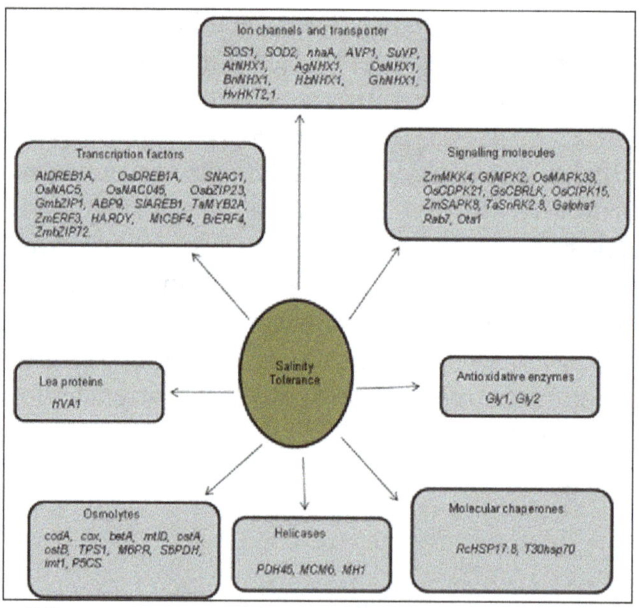

Genes coding for different transcription factors, lea proteins, osmolytes, helicases, molecular chaperones, antioxidative enzymes, signalling molecules, ion channels and transporter which are known to impart salinity tolerance to different plants when overexpressed.

Mannitol, Sorbitol and Ononitol

Bacterial gene mtlD, which codes for mannitol-1-phosphate dehydrogenase, when expressed in tobacco, causes mannitol accumulation. Similarly, ectopic expression of mtlD from E.coli in wheat plants results in 1340 enhanced tolerance to salt stress due to protective role of mannitol. Arabidopsis plants transformed with celery's mannose-6-phosphate reductase (M6PR) gene produced mannitol and were found more tolerant as comparative to wild type under salinity stress. Japanese persimmon (Diospyros kaki Thunb. cv Jiro) when transformed with apple cDNA for S6PDH encoding NADP dependent sorbitol-6-phosphate dehydrogenase, accumulates sorbitol and showed higher salinity tolerance than untransformed plants as reflected by higher ratio of variable to maximum fluorescence. Likewise improved salt and drought tolerance was found in Nicotiana tabacum, when transformed with cDNA of imt1 encoding for myo-inositol-omethyltransferase. This was indicated by accumulation of methylated inositol D-ononitol exceeding 35 mmol/g fresh weights and higher CO_2 fixation capacity in transgenic plants under stress condition.

Proline

In plants, proline is synthesized from its precursor glutamic acid and acts as an osmoprotectant under osmotic stress conditio. Two enzymes play important role in the biosynthesis of proline which are pyrroline-5-carboxylate synthase (5PCS) and pyrroline-5- carboxylate reducatse (P5CR). Transgenic rice plants of mouth bean P5CS gene encoding for pyrroline-5-carboxylate synthase under constitutive or stress inducible promotor showed significant salinity toleranc. Likewise transformed Nicotiana tabacum plants with cDNA encoding delta-1- pyrroline-5-carboxylate synthetase (P5CS) from Vigna acontifolia were found more tolerant to salinity and drought stress. Similarly, tobacco plants engineered for higher proline production by removing feedback inhibition of rate limiting enzyme in proline biosynthesis showed drought tolerance.

Engineering Plants for Transporters and Ion Channels

There is a lot of genetic diversity in plants with respect to sensitivity to NaCl. Accordingly, they are classified as halophyte (salt tolerant) and glycophyte (salt sensitive). Halophyte can grow at higher concentration of salt than glycophyte, they do it; either by, excluding Na+ or accumulating Na+ in cellular compartments like vacuoles higher K^+ /Na^+ ratio is maintained in the cytoplasm. Excess Na+ leads to the loss of ionic homeostasis. Potassium acts as a coenzyme for many cytoplasmic enzymes, but when excess Na+ is present in rhizosphere, it competes for K+ particularly at low affinity K^+ channels, leading to low K^+ /Na^+ ratio in cytoplasm. Excess Na+ in cytoplasm is equally harmful to both halophyte and glycophyte. Genetic engineering of genes for antiporter or ion channels have been successful in generation of salt tolerant plants by maintaining higher K^+ /Na^+ ratio. Overexpression of AtSOS1 encoding a plasma membrane Na^+ /H^+ antiporter which share sequence similarity to Na^+ /H^+ antiporters from bacteria and fungi leads to salt tolerance in Arabidopsis. In the same vein, plasma membrane Na^+ /H^+ antiporter gene SOD2 from Schizosaccharomyces pombe resulted in enhanced salt tolerance in Arabidopsis when overexpressed. Similarly, nhaA of E.coli which encodes for Na^+ /H^+ antiporter when expressed in rice improved salt tolerance. Another strategy of salinity tolerance by plants is to sequester Na+ ion into vacuole, so as to prevent cytosol from its toxicity. The transfer of Na^+ into vacuole is driven by a vacuolar Na^+ /H^+ antiporter which in turn is driven by the electrochemical gradient of protons generated by the vacuolar H^+ -ATPase and H+ -pyrophosphatase. The overexpression of AVP1 encoding for vacuolar H+ -pyrophosphatase in Arabidopsis results in salinity tolerance. Recently Liu have isolated and characterized a gene ScVP from Suaeda corniculata, encoding a vacuolar H^+ -pyrophosphatase (V-H $^+$ -PPase), whose ectopic expression in Arabidopsis caused salinity tolerance. Genetic engineering of cotton plants with vacuolar H^+ -pyrophosphatase (AVP1) from Arabidopsis confers salinity and drought tolerance. Apse showed that Arabidopsis thaliana plants expressing AtNHX1, a vacuolar Na^+ /H^+ antiporter, were salt tolerant. Similarly, overexpression of AtNHX1 in tomato and Brassica enhances their salt tolerance. Likewise when a vacuolar Na^+ /H^+ antiporter gene AgNHX1 from Atriplex gmelini, is overexpressed in rice, it enhanced salt tolerance. Rice OsNHX1 encoding for vacuolar Na^+ /H^+ antiporter when overexpressed, results in increased salinity tolerance. Transgenic maize and wheat for AtNHX1 showed higher tolerance to salinity. Similarly, the overexpression of vacuolar Na^+ /H^+ antiporter genes; HbNHX1 (barley), GhNHX1 (cotton) and BnNHX1 (Brassica napus) in tobacco, improved salinity tolerance. Recently, it was found that when AtNHX5 was expressed in Paper mulberry, it conferred salinity and drought tolerance in transgenic. Sodium ions enter the cell through several low and high affinity potassium carriers. There are three types of low affinity K+ transporters: inward rectifying channels (KIRC), outward rectifying channels (KORC) and voltage independent non-selective cation channels (NSCC). Arabidopsis AtHKT1 carry out circulation of Na+ in plant by mediating Na^+ loading in leaf phloem and Na^+ unloading from the root phloem sap. A mutation in AtHKT leads to increase of Na+ concentration in shoot and enhancement of plant Na+ sensitivity. Mian have shown that when barley HvHKT2;1 was overexpressed, transgenic plants were more tolerant to salt due to increased Na+ loading into xylem and accumulation of Na+ into shoot. Therefore, the increased uptake and translocation of Na^+ is also responsible for salinity tolerance.

Engineering of Antioxidative Enzymes

Reactive oxygen species (ROS) are produced under normal conditions in plants; but under stress conditions, their level is highly increased. Plants have devised antioxidative defense system to scavenge harmful ROS, and protect plant cells from oxidative injury. This antioxidative defense involves both enzymatic and non-enzymatic metabolites. Various transgenic overexpressing antioxidative enzymes like superoxide dismutase, glutathione reductase, glutathione peroxidase and ascorbate peroxidases; have been generated, which show tolerance to various abiotic stresses. Alfalfa helicase MH1 when expressed in arabidopsis enhances salinity and drought tolerance, by improving its antioxidative defense. Lots of transgenic have been produced by engineering methylglyoxal pathway. Methylglyoxal is a cytotoxic compound which accumulates to higher concentration in plants during stress conditions. Glyoxalase I and glyoxalase

II are the enzymes involved in detoxification of methylglyoxal. Many transgenic plants overexpressing genes; GlyI and GlyII encoding for enzymes glyoxalase I and glyoxalase II respectively, have been found to show salinity tolerance.

Engineering for Transcription Factors

In line with this argument that single gene level management, for stress tolerance is not so affective; efforts were made to raise transgenic plants for stress inducible transcription factors: as a transcription factor regulates many genes. It is also likely that many stress responsive genes, may share a common transcription factor. Various transcription factors belonging to the families of DREB, NAC, MYB, MYC, Cys2His2 zinc finger, bZIP, AP2/ERF and WRKY are known to be involved in salt stress tolerance. They bind to the promoter and regulatory elements of genes responsive to stress. Member s of different groups may be involved in a single response, and members of the same group may also be responsible for different kind of stress responses. Many transgenic tolerant to salinity stress have been produced through genetic engineering of gene s for transcription factors. Transgenic Arabidopsis plants overexpressing AtDREB1A were found tolerant to dehydration and freezing. Similarly, overexpression of rice OsDREB1A in Arabidopsis results in freezing dehydration and salt tolerance. Rice plants overexpressing OsDREB2A are comparatively tolerant to salinity and dehydration stress than untransformed plant. The plant specific transcription factor group NAC (NAM, ATAF1/2, and CUC2) is required for its role in plant development and stress response. Transgenic rice plants overexpressing SNAC1 (stress responsive NAC 1) showed enhanced salinity and drought tolerance. Similarly, rice plants overexpressing SNAC2 (a rice NAC transcription factor group), exhibit higher salinity tolerance. Overexpression of OsNAC5 in rice and Arabidopsis ectopic expression in enhanced salinity and drought tolerance, while knockdown of this gene in rice by RNAi lead to salt susceptibility. Likewise overexpression of ONAC045 encoding for NAC transcription factor gene in rice enhanced salinity and drought tolerance. In the same vein, OsbZIP23, a member of basic leucine zipper (bZIP) transcription factor family from rice, when overexpressed results in drought and salt tolerance. Recently a gene, GmbZIP1, encoding for a novel bZIP transcription factor from soybean was found to provide multiple abiotic stress tolerance (salt, drought and low temperature) to transgenic plants of Arabidopsis and tobacco, when overexpressed. The constitutive expression of maize ABP9; encoding a bZIP type transcription factor in Arabidopsis, results in enhancement of multiple stress tolerance including high salt, drought, freezing and oxidative stress. Ectopic expression of a maize gene ZmbZIP72 (a bZIP transcription factor) in Arabidopsis, result in salinity tolerance. Tomato plants overexpressing SlAREB1 show drought and salt tolerance. Similarly, ZFP179 (a salt responsive gene) imparts enhanced salt tolerance after it was overexpressed in rice. Likewise transfer of wheat TaMYB2A in Arabidopsis provides multiple stress tolerance including salt and drought. Zhang have shown that Ectopic expression of soybean GmERF3 gene encoding for AP2/ERF transcription factor tobacco enhances tolerance to both biotic and abiotic stresses. Transgenic Trifolium alexandrinum L. of a gene HARDY from Arabidopsis were found more tolerant to salinity and drought stress. When MtCBF4 gene from M.truncatulla encoding for a transcription factor was overexpressed in Arabidopsis resulted in enhanced salinity and drought tolerance. Similarly, transgenic Arabidopsis plants overexpressing of BrERF4 from Brassica showed enhanced salinity and drought tolerance.

Transgenic for Helicases

Helicases (RNA or DNA) are proteins involved in unwinding double stranded DNA / RNA. These ATP dependent molecules play a regulatory role in basic genetic processes including replication, transcription, translation and repair or recombination. They have been classified in five superfamilies based on their amino acid sequence, from superfaimily 1 (SF1) to superfamily 5 (SF5). Sanan-Mishra have shown that pea DNA helicase gene (PDH45), when overexpressed in tobacco enhances salinity tolerance in transgenic plants without affecting yield. Pea DNA helicase 47 (PDH47) transcripts were found induced in both shoot/root in response to salinity and cold. This purified recombinant protein showed ATP dependent DNA/RNA helicase activity and DNA dependent ATPase activity. Liu have isolated and characterized a salt inducible DEAD box helicase; AvDH1, from halophyte Apocynum venetum and suggested its possible role in salt tolerance. Similarly, ectopic expression of a DEAD box helicase (MH1) from Medicago sativa in Arabidopsis, results in salinity and drought tolerance by enhancing ROS scavenging capacity and osmotic adjustment. Chung have isolated salt inducible DEAD box helicase from soybean named GmRH and speculated its role in RNA processing under salinity and chill stress. Dang have shown that DNA helicase MCM6 transcript was upregulated in pea during salt and cold stress but not in drought or ABA treatment. Transgenic tobacco plants overexpressing MCM6 were found salinity tolerant. The investigators also reported stress responsive elements in promotor of MCM6.

Engineering of Molecular Chaperones

Molecular chaperones are a diverse group of proteins involved in various cellular functions comprising folding/unfolding, macromolecular assembly/disassembly, keeping proteins in their native state and preventing their aggregation under various stress conditions, helping in protein synthesis/degradation and targeting to their cellular compartments. Of late they have been implicated in various physiological processes and plant defense under stress conditions. Reddy have isolated pgHsc70 (encoding for cytoplasmic HSP70) from Pennisetum glaucum and suggested its probable role in plant salinity tolerance as it imparts salinity tolerance to transformed Similarly, transformed E.coli with salt inducible gene DcHsp17.7 encoding for a small heat shock protein (SHSP) from Daucus carota L. were found to show enhanced salinity tolerance. Jiang observed that E.coli, yeast and Arabidopsis transformed with RcHsp17.8 (encoding for a SHSP) from Rosa chinensis were tolerant to multiple stresses. According to Montero-Barrientos ectopic expression of Trichoderma harzianum's T30hsp70 gene in

Arabidopsis results in salt, osmotic and oxidative stress tolerance. The heat shock proteins aren't functionally limited to stress conditions, but do play role in normal development and function of plant and various cellular organelles. Constan have shown that stromal HSP100 protein is required for normal chloroplast development. The mutant of atHSP93-V which encodes for the homolog of HSP93, are smaller, paler than wild type and have chloroplasts with less thylakoid membranes.

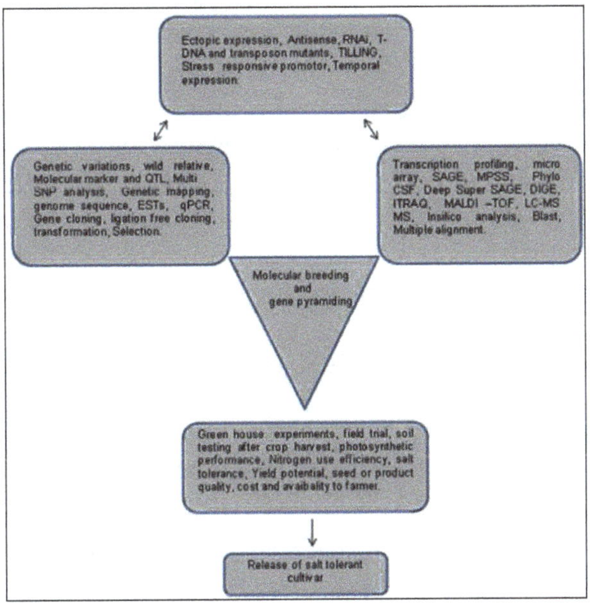

Schemataic diagram showing strategies and techniques involved for the production of salt tolerant cultivar at field level. SNP: single nucleotide polymorphism; ESTs: express sequence tags; qPCR: quantitative real time ployperase chain reaction; SAGE: serial analysis of gene expression; DIGE: differential in gel electrophoresis; iTRAQ: isobaric tag for relative and absolute quantification; MALDI-TOF: matrix assisted laser desorption ionization-time of flight; LC-MS: liquid chromatography coupled with mass spectrometry; ELISA: enzyme linked immunosorbent assay; TILLING: targeted induced local leisions in genome; RNAi: RNA interference.

Engineering for Lea Proteins

Late embryogenesis proteins are a group of hydrophilic proteins produced late during embryo development, and constitute about 4% of the total cellular proteins. These proteins have been classified into six groups; based on their amino acid sequence, mRNA homology, and expression pattern. They carry out various functions like acting as hydrating buffers, sequestering ions, helping in renaturation of proteins and acting as chemical chaperones. It was found that late embryogenesis abundant (LEA) protein gene HVA1 from Hordeum vulgare L. upon transformation into rice confers salinity and drought tolerance to transgenic plants. The stress tolerant features (including salt and drought tolerance) of HVA1 gene transformed plants have further been proven in Basmati rice and Morus indica. In the same vein, ectopic expression of PM2 from soybean; encoding for a type 3 LEA protein in E.coli, results in salinity tolerance and therefore suggesting its probable role in salinity tolerance in plants.

Engineering Plants for Signaling Molecules

Exposure of any stress to plants causing change in normal plant development is perceived by some kinds of sensor which leads to a signaling cascade resulting in a stress response by plant. These signaling pathways may be specific or non-specific depending upon the type of stress. Signaling pathways for different stresses may crosstalk. These signaling pathways involve many signaling molecules, and may be ABA dependent or independent. Ca^{2+} is most common secondary messenger in plants responding to various stimuli. Ca^{2+} is known to be involved in most common pathway for salt stress; Salt Overly Sensitive (SOS) pathway. A change in cytoplasmic Ca^{2+} transient is sensed by SOS3; a calcium binding protein. This in the presence of Ca^{2+} activates SOS2; a serine-threonine protein kinase. SOS3-SOS2 kinase complex regulate the expression, as well as activity of Na+ /H+ exchanger, which is responsible for salinity tolerance.

Mitogen activated protein kinases (MAPK) are known to be involved in signaling of multiple abiotic stress including salt, drought, temperature, and other physiological processes like cell division. MAPK cascade involve three tiers of protein kinases: MAPK, MAPK kinase (MAPKK) and MAPKkinase kinase (MAPKKK). MAPK is activated by MAPKK (MAPK kinase) by phosphorylation at two residues and MAPKK in turn is activated by MAPKKK. Many transgenic plants have been produced with high salinity tolerance by engineering MAPK cascade. Ectopic expression of Nicotiana protein kinase MAPKKK/NPK1 in maize, leads to the activation of oxidative signal eventually enhancing salt, heat and cold tolerance. A novel MAPKK from maize, ZmMKK4, when overexpressed in arabidopsis, confers salinity and drought tolerance. Zhang have shown that a MAPK from cotton GhMPK2 is induced by salt, ABA and drought stress. Its overexpression into tobacco leads to salinity and drought tolerance. In another study it was found that rice plants overexpressing OsMAPK33 are more sensitive to salt stress than wild type. There are also calcium dependent protein kinases (CDPKs), which are involved in salt stress response. Asano have characterized one CDPK gene OsCDPK21 from rice which when was overexpressed, results in enhanced salinity tolerance in transgenic.

Ectopic expression of GsCBRLK gene; encoding plant specific calcium-dependent calmodulin binding receptor like kinase from Glycine soja in Arabidopsis, enhanced salt and ABA tolerance. Another signaling molecule, Calcineurin B-like (CBL) proteins, are a group of Ca^{2+} sensor in plants. They play an important role in relaying the signal in diverse stress response by interacting with CBL-interacting protein kinases (CIPKs). Xiang have characterized a number of OsCIPKs genes in rice, for their stress inducibility and stress tolerance, out of those OsCIPK15 was responsible for salinity tolerance in transgenic when was overexpressed. Another family of kinases which play important role in signaling stress response belongs to; sucrose non-fermenting 1-related protein kinase 2 ($SnRK_2$) families. Recently a gene from maize, ZmSAPK8 of SnRK2 family, has been cloned. The overexpression of this gene into Arabidopsis confers salinity tolerance, along with upregulation of transcription of other stress marker genes like RD29A, RD29B, RAB18, P5CS1, ABI1 and DREB2A. Similarly, in an another study it was found that ectopic expression of wheat TaSnRK2.8 of SnRK2 family, in Arabidopsis improves salinity tolerance along with upregulation of transcripts of ABA biosynthesis genes. Plant lectin receptor like kinases (LecRLKs) are also known to mediate signaling during stress response. G-Protein coupled receptors (GPCRs) are known to perceive extracellular signals, and transduce subsequently to heteromeric G-proteins, which further pass signal to downstream effector. When gene from Pisum sativum (Galpha1); encoding for G-alpha subunit, was overexpressed in Arabidopsis, it enhanced salinity and heat tolerance. Ectopic expression of Rab7 from pennisetum glaucum; encoding for Rab-GTPase (a GTP binding protein), in tobacco enhanced salinity tolerance. Conti) have identified two SUMO proteases; OVERLEY TOLERANT TO SALT 1 (OTS1) and OTS_2, in Arabidopsis. Double mutant of ots1 and ots2 is salt sensitive; whereas, overexpression of ots1 confers salinity tolerance to the transgenic.

References

- Siebert, S; Döll, P; Hoogeveen, J; Faures, J M; Frenken, K; Feick, S (2005). "Development and validation of the global map of irrigation areas". Hydrol. Earth Syst. Sci. 9 (5): 535–47. doi:10.5194/hess-9-535-2005
- Salinity-tolerance-in-plants-Breeding-and-genetic-engineering-230806686: researchgate.net, Retrieved 14 March, 2020
- Rao, V.S. (2000). Principles of Weed Science. Enfield, New Hampshire: Science Publishers, Inc. pp. 39–48. ISBN 1-57808-069-X
- "A Landscaper's Guide to Mulch" (PDF). California Integrated Waste Management Board. Archived from the original (PDF) on 2011-06-01. Retrieved 12 Feb 2012
- Turgeon; et al. (2009). Weed Control in Turf and Ornamentals. Upper Saddle River, New Jersey: Pearson Education, Inc. pp. 127–129. ISBN 0-13-159122-3

Genetically Modified Plants

5

- Genetically Modified Rice
- Golden Rice
- Genetically Modified Wheat
- Genetically Modified Tomato
- Genetically Modified Maize

The plants whose DNA are modified through genetic engineering for developing new traits in crops are termed as genetically modified plants. It includes golden rice. This chapter sheds light on various genetically modified plants for an in-depth understanding of the subject.

Genetically modified (GM) plants, also called transgenic plants, are designed to acquire useful quality attributes such as insect resistance, herbicide tolerance, abiotic stress tolerance, disease resistance, high nutritional quality, high yield potential, delayed ripening, enhanced ornamental value, male sterility, and production of edible vaccines.

Another major goal for raising the GM plants is their application as bioreactors for the production of nutraceuticals, therapeutic agents, antigens, monoclonal antibody fragments biopolymers, and so forth. Thus, GM plants can potentially affect many aspects of modern society, including agricultural production and medical treatment. Despite these potential applications, the use of GM plants for human welfare has been restricted owing to various concerns raised by the public and the critics. These concerns are divided into different categories, namely, health, nutritional, environmental, ecological, socioeconomic, and ethical concerns. These concerns include those arising due to properties of GM plants themselves, those resulting from the spread of the transgenes to other organisms, and also those resulting from their release into the environment.

Concerns Related to Health and Nutritional Status

In case the products of GM plants are to be consumed by humans and animals, there is always a fear and risk in the society that these plants may create health problems or may lead to the development of newer microbial strains that may be pathogenic. Further, the plants themselves may be susceptible to such risks. Public and critics are also skeptical about the nutritional content and quality of the GM plants.

Susceptibility to Allergens

One of the major distressing problems with nontraditional proteins in GM foods is the risk of introducing allergens (usually glycoproteins) into the food supply of humans and animals. The public is concerned about the nature of these new food proteins as their allergenic or nonallergenic qualities are unknown. Allergenicity has been demonstrated in transgenic soybeans due to the transfer of a major food allergen from Brazil nuts. On the other hand, the scientists believe that the food allergens are found only in a few defined sources (peanut and other grain legumes, shellfish, tree nuts, etc.), and hence, only a dozen foods may produce allergic reactions. Moreover, allergenicity occurs when these food allergens are present in large proportions in the food and the individuals are sensitized to them over time to cause any adverse effects. Thus, it is highly unlikely for new allergens to be introduced into the food supply from GM plants.

Transfer of Antibiotic Resistance Gene to Microbes and Reduced Efficacy of Antibiotic Therapy

Public is also concerned about the potential risks associated with gene transfer from plants to microbes. It is speculated that the consumption of GM foods containing antibiotic resistance marker gene (e.g., *Npt* II gene encoding neomycin phosphotransferase for resistance to kanamycin and neomycin or gene encoding β-lactamase for resistance to ampicillin) by humans and animals may lead to transfer of these genes from GM food to microflora in the gut of humans and animals or to the pathogens in the environment transforming them into strains that are resistant to antibiotic therapy. The transfer of antibiotic resistance gene to unrelated microorganisms such as *Aspergillus niger* has also been demonstrated. Biotechnology scientists, however, are of the opinion that the *Npt* II gene used to develop GM plants currently in the market is safe for use because there is no evidence of allergenicity or toxicity related to it.

Moreover, humans are also susceptible to consuming several kanamycin resistant bacteria that occur naturally in the environment. Human gut is reported to contain 10^{12} kanamycin-resistant bacteria and by consuming a tomato harboring *Npt* II gene, the increase in frequency of kanamycin-resistant bacteria in the gut amounts only to 10^{-6} %. Furthermore, acid conditions prevalent in stomach or rumen inactivate or degrade the encoded enzyme, neomycin phosphotransferase II. Also neomycin phosphotransferase II requires ATP for its activity, which is present in extremely low concentrations in the gut. Regarding the use of gene for selection of bacterial recombinants, it is not transferred to plants. Moreover, the gene is considered safe because it does not encode for any product in plants. The *Npt* II and genes have been declared safe to use in GM plants. The public is, however, reluctant to accept this fact. Looking to the views of public, scientists have also developed nonresistance based selectable marker genes such as green fluorescent protein encoding gene (*Gfp*) andβ-glucuronidase gene (*Uid* A).

Besides, intron-containing *Npt* II gene has also been assessed as an efficient selectable marker in plant transformation. Due to insertion of intron in the *Npt* II gene, the theoretical risk of gene flow from GM plants to enteric bacteria is eliminated. Strategies for the removal of antibiotic resistance genes have also been devised. One such strategy is the cloning of selectable marker gene and the transgene on two separate transfer DNA (T-DNA) molecules in a single plasmid or on two separate plasmids that are contained in one or more *Agrobacterium tumefaciens* strains used for plant transformation. The transgene and selectable marker gene are, thus, inserted at the loci, which should recombine at reasonably high frequencies so that the transgene can be segregated from the selectable marker gene in the next generation. Second strategy to eliminate the selectable marker gene is to flank it with direct repeats of recognition sites for a site-specific recombinase so that the marker gene can be easily excised from the plant genome by recombinase-mediated site-specific recombination. Examples included in this category are the Cre/*lox* recombination system of bacteriophage P1, Flp/*frt* recombination system of yeast 2 μm plasmid and R/*Rs* system of *Zygosaccharomyces rouxii.*

A common feature of these systems is that the first round of transformation produces transgenic plants with the selection marker between two directly oriented recognition sites for the respective recombinase. After expression of recombinase, either by crossing in plants expressing the enzyme, by transient expression via second transformation, or by the use of an inducible promoter, the recombinase reaction is initiated resulting in marker-free transgenic plants. Marker gene may also be eliminated by placing it on a transposable element resulting in its loss after transposition. The transgene by itself may be mobile and the activation of transposase allows the relocation of the desired transgene to a new chromosomal position. Genetic crosses and segregation may dissociate the two transgenes. Another novel strategy for the production of marker-free GM plants involves DNA deletion based on intrachromosomal homologous recombination between two homologous sequences, for example, by incorporating *att*sequence of λ bacteriophage.

Development of New-line Microbial Strains

The third health risk is related to the ability of GM plants to create new toxic organisms. It is speculated that some nonpest microbial strains may acquire pathogenic trait by gene flow from GM plants. The risk can also be a new host being infected by a virus or recombining to form a more deadly virulent virus. Some plant pathologists also hypothesize that development of virus-resistant plants may allow viruses to infect new hosts through transencapsidation. Virus-resistant plants may also lead to the creation of new viruses through an exchange of genetic material or recombination between RNA virus genomes. Another matter of concern is that a small fraction of the DNA released from GM plants into soil may bind to the clay particles and hence protected from degradation. It is speculated that the soil bacteria may undergo transformation with the exogenous DNA of GM plant. This is, however, a rare possibility as the amount of DNA derived from GM plants as a proportion of the total DNA in the soil is likely to be very small, even if such plants are grown on a commercial scale. Moreover, the longevity of DNA in soil depends on various factors, including soil type and the presence of deoxyribonucleases in soil. Laboratory microcosm experiments have shown all but 0.1% of the target DNA from transgenic tobacco plants gets degraded within 40 days.

Environmental and Ecological Concerns

Large-scale cultivation of GM plants expressing viral and bacterial genes and their release into the environment is considered to be a threat and called as "genetic pollution" by the critics. The risk of a transgene spreading in the environment is related to the likelihood for out-crossing, horizontal gene transfer, and the phenotype imparted by the gene. Debates about the commercial introduction of GM plants in some parts of the world have led to questions about their potential impact on the environment unless necessary safeguards are taken into account. Various concerns that have arisen due to the application and release of GM plants into the environment are given in this section. It should, however, be acknowledged that agriculture inevitably has an impact on the environment and these concerns are not specific for GM plants.

Transgene Escape to Wild-type Plants

There is a potential risk that the GM plants may hybridize (or cross-breed) with sexually compatible wild-type species. This genetic exchange is possible due to wind pollination, biotic pollination or seed dispersal. This may have an impact on the environment through the production of hybrids and their progeny. In an example, virus-resistant squash commercialized in 1994 was demonstrated to transfer its virus resistance gene to wild squash (*Cucurbita pepo*), an agricultural weed native to the southern United States, thereby decreasing its value to squash breeders. On the other hand, it is significant to note that for an effective pollen transfer to occur, the GM plants must be close enough to the wild species, should flower at same time, and must be genetically compatible. Further, the risk of any gene transfer to related weedy species through pollen has been eliminated by devising chloroplast transformation procedures. This is because, in many crop species, chloroplasts display only maternal inheritance.

Selective Advantage to GM Plants in Natural Environments and Generation of Superweeds

The concern of gene flow from GM plants to weedy relatives via pollination is quite intense. It is considered that the transfer of encoded characteristics to weed species could potentially give them a selective advantage, consequently leading to the generation of "superweeds". Moreover, the newly introduced traits may make a plant, especially herbicide tolerant plant, more persistent or invasive (weedy) in agricultural habitats. It is, however, pertinent to note that the risk of gene transfer to weeds is similar with both conventional and GM plants and is not contingent on how these genes have been introduced into plants. Such a risk of gene flow has always existed since the advent of modern plant breeding, even when there were no GM plants, and this can occur where possible. Several studies have demonstrated that tolerance to particular herbicide is often more likely to develop by evolution from within the weed gene pool rather than by gene flow from herbicide-tolerant plants. Nevertheless, the current scientific evidence indicates that the weediness arises from many different characters and that the addition of one gene is unlikely to cause a crop to become a weed. The transfer of novel genes from transgenics (or even conventionally bred plants) to weeds depends on the nature of the novel gene and the biology and ecology of the recipient weed species. The probability of successful out-crossing thus depends on sexual compatibility, physical proximity, distance of pollen movement both out of and into the GM plants, and ecology of recipient species. Thus, only a few plants such as oilseed rape, barley, wheat, beans, and sugar beet can hybridize with weeds. For example, oilseed rape has been reported to hybridize with hoary mustard, wild radish, and other wild *Brassica* species. Furthermore, the transfer of herbicide tolerance gene is unlikely to confer any competitive advantage to hybrids outside agricultural areas. It is also comforting to recognize that there is no proven evidence of enhanced persistence or invasiveness of GM plants and no major superweeds have developed so far.

Effect on Nutritional Composition of Plants

It is also speculated that the nutritional composition of GM products may be affected in GM plants. Another concern is that the transgenes from animals (obtained from fishes, mouse, human, and microbes) introduced into GM plant for molecular farming may pose a risk of changing the fundamental nature of vegetables. In a study, it was reported that as compared to non-GM soybean, GM plants exhibited lower levels of isoflavones. This finding also raised a doubt on the regulatory system for the release of the GM plants. However, later it was found that the concentration of isoflavone in GM soybean was within the normal range.

Mixing Genes from Unrelated Species

The public is worried about the risk that the GM plants can spread through nature and interbreed with natural organisms, thereby contaminating "non-GM" environments. This would in turn affect the future generations in an unforeseeable and uncontrollable way. Such worries, however, ignore the history of plant breeding and the existing overwhelming sequence similarity of genes across kingdoms.

Development of Tolerance to Target Herbicide

It is viewed that the repeated use of the same herbicide in the same area to remove weeds amongst genetically modified herbicide-resistant crops (HRCs) (tolerant to single herbicide) will exacerbate the problem of herbicide-tolerant weeds. Another matter of concern relates to the plants carrying different herbicide tolerance genes to become multiply tolerant to several herbicides by pollination between adjacent plants. In several closely studied examples in Canada, farmers have detected oilseed rape plants tolerant to three different herbicides (note that two were acquired from GM plants and the third possibly from conventional breeding). The development of multiple tolerances in "volunteer" crop plants (from seeds remaining viable in agricultural soil) may also exert an impact on the environment by necessitating the use of less environment-friendly and possibly outdated herbicides by the farmers. On the other hand, the proponents believe that herbicide resistance develops due to excessive application of herbicide and is not exclusively associated with gene transfer from genetically modified HRCs. Thus, the pressure on weeds to evolve resistant biotypes has been reported to be pronounced with the excessive application of herbicides such as glyphosate, sulphonylureas, and imidazolinones.

Sustainable Resistance in Insect Pests

It is possible that the widespread use of disease-resistant GM plants may lead to the evolution of several insect pests that are resistant to pesticides. For example, *Bt* crops may develop resistance to *Bt* biopesticide, a permitted biopesticide successfully used by organic farmers in the integrated pest management (IPM) programs. There is to date no reported evidence of insect resistance to *Bt* crops under field conditions although *Bt* resistant insects (e.g., cotton budworm and bollworm) have been observed in areas where *Bt* biopesticides are sprayed on crops. It has been a matter of concern that the development of such resistance may lead to the loss of the potential of the *Bt*-biopesticide, which may in turn make it necessary for organic farmers to resort to less environmentally acceptable chemical pesticides. Therefore, proper resistance management strategies along with this comparatively newer technology are imperative. The most widely used is the 'high-dose refuge' strategy designed to prevent or delay the emergence of Bt toxin-resistant insects. Scientists are of the opinion that this strategy should be followed without fail, as the rate of noncompliance can increase the risk of plant resistance breakdown.

Harm to Nontarget Organisms

Nontarget effect, that is, undesirable effect of a novel gene (usually conferring pest or disease resistance) on "friendly" organisms in the environment, is another concern related to GM plants. As many nontarget microbes harbor on plant surfaces or some insects harbor on flowers, it becomes quite challenging to target the insect resistance gene product to appropriate plant tissues and hence kill pests without exerting any adverse effect on friendly organisms such as pollinators and biological control agents. This is particularly difficult where the benign or beneficial organism is related and physiologically similar to the pest to be targeted. One of the most significant studies of nontarget impacts of GM plants has been the killing of monarch butterfly in the United States by Bt insecticidal proteins. It should, however, be noted that the pesticidal sprays used on *Bt* or non-*Bt* corn may be more harmful to the monarch butterfly as compared to *Bt* corn pollen. Thus, in evaluating the use of *Bt* crops and the possible environmental damage caused, it is important to take into account the environmental damage caused by the use of pesticides in agriculture generally. It is argued that millions of birds and billions of insects, both harmful and beneficial, are killed each year due to excessive use of pesticides. It is, however, suggested that the scale and pattern of use may mitigate the effects of *Bt* on nontarget populations. Furthermore, when toxins are produced within plant tissues, nontarget organisms are exposed to a much lesser extent than with spray applications because only those organisms which feed on the plant tissues come into contact with the toxin.

Harmful effect of *Bt* toxin residues in the soil after harvest of the GM crop on soil invertebrates has been another matter of concern. An investigation of the effect of *Cry1Ab* released from the roots and crop residues on soil organisms revealed the presence of toxin in the guts and casts of tested earthworms. There was, however, no significant difference in their mortality or weight. Moreover, no difference in the total number of other soil organisms (including nematodes, protozoa, bacteria, and fungi) between the soil rhizosphere of *Bt* and non-*Bt* crops was detected.

Increased use of Chemicals in Agriculture

On one hand, the transgenes conferring herbicide resistance have been criticized because these would maintain, if not promote, the use of herbicides and their attendant problems. Similarly, there is a concern that the insect-resistant and disease-resistant GM plants will increase the application of insecticides and pesticides, respectively. On the contrary, reports demonstrate that there is no significant change in the overall amount of herbicide use in the United States since the introduction of GM soybeans. An analysis by soybean growers at the United States has shown that $7.2 millions of other herbicides were replaced by $5.4 millions of glyphosate. This substitution, thus, resulted in the replacement of highly toxic and more persistent herbicides with that of glyphosate. Furthermore, it has been reported that herbicide-tolerant oilseed rape eliminates the use of >6,000 tons of herbicide in the growing season.

Loss of Biodiversity

The public has long been worried about the loss of plant biodiversity due to global industrialization, urbanization, and the popularity of conventionally-bred high-yielding varieties. It is speculated that the biodiversity will be further threatened due to the encouraging use of GM plants. This is because development of GM plants may favor monocultures, that is, plants of a single kind, which are best suitable for one or other conditions or produce one product. Further, the transformation of more natural ecosystems into agricultural lands for planting GM plants is adding to this ecological instability.

Another point of concern is the loss of weed diversity that may occur due to gene flow from HRCs to weeds. It is argued that because the currently available HRCs confer tolerance to broad-spectrum herbicides such as glufosinate and glyphosate, their extensive use may shift the diversity of weeds in agricultural habitats. However, weeds exhibit considerable plasticity and adapt to a wide range of cultivation practices. Experience with conventional agriculture has shown that weed species composition varies within the same crop among different fields and at different times of year. Thus, weed population shifts are natural ecological phenomena in crop management and should not be viewed as exclusive to GM plants.

Unpredictable Gene Expression

It is speculated that the random gene insertion, transgene instability, and genomic disruption due to gene transfer may result in unpredictable gene expression. Such a risk is, however, unlikely to be unique to GM plants or of any significance considering our current knowledge of genomic flux in plants.

Alteration in Evolutionary Pattern

Plants adapt to the fluctuations in the environment through changing their genes and developing better races called "evolved races". These mutations, however, occur at a very low frequency (i.e., one in about 10^9/gene/generation). It is hypothesized that the cultivation of GM plants by the farmers at an increasing rate throughout the world may change the evolutionary pattern drastically. Another concern is the evolution of non-GM plants through hybridization with GM plants.

Loss of Ecosystem in Marginal Lands

As new plants are introduced mainly to marginal lands, loss of natural ecosystems in these areas has also been a matter of concern.

Contamination of Soil and Water

It is also sometimes argued that the widespread introduction of HRCs will increase the use of herbicides, which will in turn contribute to the contamination of soil and ground water. However, this is not the case. The cultivation of HRCs in the United States has been reported to facilitate zero-till agronomic system, which contributes to a reduction in soil erosion. The release of *Bt* toxin into the soil after harvest of *Bt* crops is also viewed as a risk factor associated with the cultivation of *Bt* crops. It has been found that *Bt* toxins remain active in soil; however, it is not necessarily an environmental hazard because *Bt* toxins must be ingested and affect only selected groups of insects. Moreover, the potential leaching rate of *Bt* toxin is reduced due to its binding and adsorption on clay particles.

Genetically Modified Rice

Genetically modified rice are rice strains that have been genetically modified (also called genetic engineering). Rice plants have been modified to increase micronutrients such as vitamin A, accelerate photosynthesis, tolerate herbicides, resist pests, increase grain size, generate nutrients, flavours or produce human proteins.

The natural movement of genes across species, often called horizontal gene transfer or lateral gene transfer, can also occur with rice through gene transfer mediated by natural vectors. Transgenic events between rice and *Setaria* millet have been identified. The cultivation and use of genetically modified varieties of rice remains controversial and is not approved in some countries.

Rice plants being used for genetic modification.

In 2000, the first two GM rice varieties both with herbicide-resistance, called LLRice60 and LLRice62, were approved in the United States. Later, these and other types of herbicide-resistant GM rice were approved in Canada, Australia, Mexico and Colombia. However, none of these approvals triggered commercialization. Reuters reported in 2009 that China had granted biosafety approval to GM rice with pest resistance, but that strain was not commercialized. As of December 2012 GM rice was not widely available for production or consumption. Advocates claim that since rice is a staple crop across the world, improvements have potential to alleviate hunger, malnutrition and poverty.

Traits

Herbicide Resistance

In 2000-2001 Monsanto researched adding glyphosate tolerance to rice but did not attempt to bring a variety to market. Bayer's line of herbicide resistant rice is known as LibertyLink. LibertyLink rice is resistant to glufosinate (the active chemical in Liberty herbicide). Bayer CropScience is attempting to get their latest variety (LL62) approved for use in the EU. The strain is approved for use in the US but is not in large-scale use. Clearfield rice was bred by selection from variations created in environments known to cause accelerated rates of mutations. This variety tolerates imidazole herbicides. It was bred by traditional breeding techniques that are not considered to be genetic engineering. Clearfield is also crossbred with higher yielding varieties to produce an overall hardier plant.

Nutritional Value

Golden Rice grains (right) compared to regular rice grains (left).

Golden rice with higher concentrations of Vitamin A was originally created by Ingo Potrykus and his team. This genetically modified rice is capable of producing beta-carotene in the endosperm (grain) which is a precursor for vitamin A. Syngenta was involved in the early development of Golden Rice and held some intellectual property that it donated to non-profit groups including the International Rice Research Institute (IRRI) to develop on a non-profit basis. The scientific details of the rice were first published in *Science Magazine* in 2000.

The World Health Organization stated that iron deficiency affects 30% of the world's population. Research scientists from the Australian Centre for Plant Functional

Genomics (ACPFG) and IRRI to are working to increase the amount of iron in rice. They have modified three populations of rice by over expressing the genes OsNAS1, OsNAS2 or OsNAS3. The research team found that nicotianamine, iron, and zinc concentration levels increased in all three populations relative to controls.

Golden Rice plants being grown in greenhouse.

Pest Resistance

BT rice is modified to express the cryIA(b) gene of the *Bacillus thuringiensis* bacterium. The gene confers resistance to a variety of pests including the rice borer through the production of endotoxins. The Chinese Government is doing field trials on insect resistant cultivars. The benefit of BT rice is that farmers do not need to spray their crops with pesticides to control fungal, viral, or bacterial pathogens. Conventional rice is sprayed three to four times per growing season to control pests. Other benefits include increased yield and revenue from crop cultivation. China approved the rice for large-scale use as of 2009.

Allergy Resistance

Researchers in Japan are attempting to develop hypoallergenic rice cultivars. Researchers are trying to repress the formation of allergen AS-Albumin.

Japanese researchers tested genetically modified rice on macaque monkeys that would prevent allergies to cedar pollen, which causes hay fever. Cedar allergy symptoms include itchy eyes, sneezing and other serious allergic reactions. The modified rice contains seven proteins from cedar pollen (7Crp) to block these symptoms by inducing oral tolerance. Takaiwa is conducting human clinical trials with this 7Crp protein as an oral vaccine.

C4 Photosynthesis

In 2015 a consortium of 12 laboratories in eight countries developed a cultivar that displayed a rudimentary form of C4 photosynthesis (C4P) to boost growth by capturing carbon dioxide and concentrated it in specialized leaf cells. C4P is the reason corn and sugarcane grow so rapidly. Engineering C4 photosynthesis into rice could increase yields per hectare by roughly 50 percent. The current cultivar still relies primarily on C3 photosynthesis. To get them to completely adopt C4P, the plants must produce specialized cells in a precise arrangement: one set of cells to capture the carbon dioxide and to surround other cells that concentrate it. Some (possibly dozens of) genes involved in producing these cells remain to be identified. Other C3P crops that could exploit such knowledge include wheat, potatoes, tomatoes, apples and soybeans.

Production of Recombinant Proteins

Human serum albumin (HSA) is a blood protein in human blood plasma. It is used to treat severe burns, liver cirrhosis and hemorrhagic shock. It is also used in donated blood and is in short supply around the world. In China, scientists modified brown rice as a cost-effective way to produce HSA protein. The Chinese scientists put recombinant HSA protein promoters into 25 rice plants using *Agrobacterium*. Out of the 25 plants, nine contained the HSA protein. The genetically modified brown rice makes the same amino acid sequence as HSA. They called this protein *Oryza sativa* recombinant HSA (OsrHSA). The modified rice was transparent. OsrHSA was soon sold to replace cow albumin for growing cells. Clinical trials were started in China in 2017, and in the US in 2019. The same Oryzogen company makes other recombinant human proteins from rice.

Ventria Bioscience uses a proprietary system known as Express Tec for producing recombinant human proteins in rice grains. Their most notable variety produces human Lactoferrin and Lysozyme. These two proteins are produced naturally in human breast milk and are used globally in infant formula and rehydration products.

Golden Rice

Golden rice is a variety of rice (*Oryza sativa*) produced through genetic engineering to biosynthesize beta-carotene, a precursor of vitamin A, in the edible parts of rice. It is intended to produce a fortified food to be grown and consumed in areas with a shortage of dietary vitamin A, a deficiency which each year is estimated to kill 670,000 children under the age of 5 and cause an additional 500,000 cases of irreversible childhood blindness. Rice is a staple food crop for over half of the world's population, providing 30–72% of the energy intake for people in Asian countries, and becoming an effective crop for targeting vitamin deficiencies.

Golden rice differs from its parental strain by the addition of three beta-carotene biosynthesis genes. The parental strain can naturally produce beta-carotene in its leaves, where it is involved in photosynthesis. However, the plant does not normally produce the pigment in the endosperm, where photosynthesis does not occur. Golden rice has met significant opposition from environmental and anti-globalization activists. A study in the Philippines is aimed to evaluate the performance of golden rice, if it can be planted, grown and harvested like other rice varieties, and whether golden rice poses risk to human health. There has been little research on how well the beta-carotene will hold up when stored for long periods between harvest seasons, or when cooked using traditional methods.

Golden rice (right) compared to white rice (left).

Genetics

Golden rice was created by transforming rice with two beta-carotene biosynthesis genes:

- Psy (phytoene synthase) from daffodil ('Narcissus pseudonarcissus').
- Crti (phytoene desaturase) from the soil bacterium Erwinia uredovora.

(The insertion of a *lcy* (lycopene cyclase) gene was thought to be needed, but further research showed it is already produced in wild-type rice endosperm.)

The *psy* and *crtI* genes were transferred into the rice nuclear genome and placed under the control of an endosperm-specific promoter, so that they are only expressed in the endosperm. The exogenous *lcy* gene has a transit peptide sequence attached, so it is targeted to the plastid, where geranylgeranyl diphosphate is formed. The bacterial *crtI* gene was an important inclusion to complete the pathway, since it can catalyze multiple steps in the synthesis of carotenoids up to lycopene, while these steps require more than one enzyme in plants. The end product of the engineered pathway is lycopene, but if the plant accumulated lycopene, the rice would be red. Recent analysis has shown the plant's endogenous enzymes process the lycopene to beta-carotene in the endosperm, giving the rice the distinctive yellow color for which it is named. The original golden rice was called SGR1, and under greenhouse conditions it produced 1.6 µg/g of carotenoids.

Vitamin A Deficiency

The research that led to golden rice was conducted with the goal of helping children who suffer from vitamin A deficiency (VAD). In 2005, 190 million children and 19 million pregnant women, in 122 countries, were estimated to be affected by VAD. VAD is responsible for 1–2 million deaths, 500,000 cases of irreversible blindness and millions of cases of xerophthalmia annually. Children and pregnant women are at highest risk. Vitamin A is supplemented orally and by injection in areas where the diet is deficient in vitamin A.

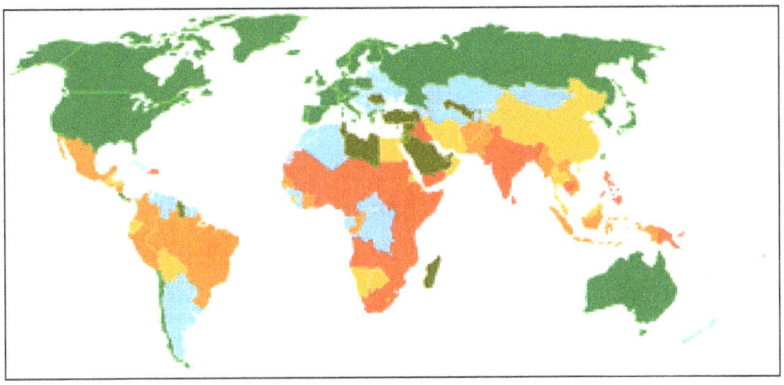

Prevalence of vitamin A deficiency. Red is most severe (clinical), green least severe. Countries not reporting data are coded blue.

As of 1999, 43 countries had vitamin A supplementation programs for children under 5; in 10 of these countries, two high dose supplements are available per year, which, according to UNICEF, could effectively eliminate VAD. However, UNICEF and a number of NGOs involved in supplementation note more frequent low-dose supplementation is preferable.

Because many children in VAD-affected countries rely on rice as a staple food, genetic modification to make rice produce the vitamin A precursor beta-carotene was seen as a simple and less expensive alternative to ongoing vitamin supplements or an increase in the consumption of green vegetables or animal products. Initial analyses of the potential nutritional benefits of golden rice suggested consumption of golden rice would not eliminate the problems of vitamin A deficiency, but could complement other supplementation. Golden Rice 2 contains sufficient provitamin A to provide the entire dietary requirement via daily consumption of some 75g per day.

Since carotenes are hydrophobic, sufficient fat must be present in the diet for golden rice (or most other vitamin A supplements) to alleviate vitamin A deficiency. Vitamin A deficiency is usually coupled to an unbalanced diet. Moreover, this claim referred to an early cultivar of golden rice; one bowl of the latest version provides 60% of RDA for healthy children. The RDA levels advocated in developed countries are far in excess of the amounts needed to prevent blindness.

Distribution

A recommendation was made that golden rice to be distributed free to subsistence farmers. Free licenses for developing countries were granted quickly due to the positive publicity that golden rice received, particularly in *Time* magazine in July 2000.

Monsanto Company was one of the companies to grant free licences for related patents owned by the company. The cutoff between humanitarian and commercial use was set at US$10,000. Therefore, as long as a farmer or subsequent user of golden rice genetics would not make more than $10,000 per year, no royalties would need to be paid. In addition, farmers would be permitted to keep and replant seed.

Genetically Modified Wheat

Genetically modified wheat is wheat that has been genetically engineered by the direct manipulation of its genome using biotechnology. As of 2017, no GM wheat is grown commercially, although many field tests have been conducted.

Wheat is a natural hybrid derived from interspecies breeding. It is theorized that wheat's ancestors (*Triticum monococcum, Aegilops speltoides*, and *Aegilops tauschii*, all diploid grasses) hybridized naturally over millennia somewhere in West Asia, to create natural polyploid hybrids, the best known of which are common wheat and durum wheat.

Wheat (*Triticum* spp.) is an important domesticated grass used worldwide for food. Its evolution has been influenced by human intervention since the dawn of agriculture.

Interspecies transfer of genes continued to occur in farmers fields during the shift from the Paleolithic diet to the diet adopted by humans following the Neolithic Revolution, or first green revolution. During the transition from a hunter-gatherer social structure to more agrarian societies, humans began to cultivate wheat and further transform it for their needs. Thus, the social and cultural roots of humans and the development of wheat have intertwined since before recorded history.

This process resulted in various wheat species that are grown for specific purposes and climates. In 1873 Wilson cross-pollinated rye and wheat to create triticale. Further transformations using cytogenic hybridization techniques enabled Norman Borlaug, father of the second Green Revolution, to develop wheat species (the semidwarf varieties) that would grow in harsh environments.

Recombinant DNA techniques were developed in the 1980s, work began on creating the first transgenic wheat, coincident with the third Green Revolution. Of the three most important cereals in the world (corn, rice and wheat), wheat was the last to be transformed by transgenic, biolistic methods in 1992, and by Agrobacterium methods in 1997. Unlike corn and rice, its widespread use in the human diet has faced cultural resistance.

Field Trials and Approvals

As of 2013, 34 field trials of GM wheat have taken place in Europe and 419 have taken place in the US. Modifications tested include those to create resistance to herbicides, create resistance to insects and to fungal pathogens (especially fusarium) and viruses, tolerance to drought and resistance to salinity and heat, increased and decreased content of glutenin, improved nutrition (higher protein content, increased heat stability of the enzyme phytase, increased content of water-soluble dietary fiber, increased lysine content), improved qualities for use as biofuel feedstock, production of drugs via pharming and yield increases.

As of December 2017, no GM wheat had been approved for release anywhere in the world.

Monsanto's MON 71800

The transgenic wheat that was furthest developed was Monsanto's MON 71800, which is glyphosate-resistant via a CP4/maize EPSPS gene. Monsanto received approval from the FDA for its use in food, but withdrew its EPA application in 2004, so the product was never marketed. It also received approval for use as food in Colombia.

Studies conducted by Monsanto showed that its nutritional components are equivalent to nontransgenic commercially available wheat, and animal studies that have used MON 71800 for feed have confirmed this. Environmental Risk assessments have been conducted by Monsanto, and government regulatory agencies have approved its use in food;

However, farmers were worried about the potential loss of markets in Europe and Asia due to public refusal of the end-product, so Monsanto withdrew its EPA application for Roundup-Ready Wheat.

Bioceres's HB_4

The Argentina-based company Bioceres developed a genetically modified wheat variety claiming to withstand drought while providing farmers with high yields. But as of 2019 this GM variety isn't approved yet because of risks for trade: export markets may not accept it.

Escape of GM Wheat Seed

In 1999 scientists in Thailand claimed they discovered glyphosate-resistant wheat in a grain shipment from the Pacific Northwest of the United States, even though transgenic wheat had never been approved for sale and was only ever grown in test plots. No one could explain how the transgenic wheat got into the food supply.

In May 2013 a strain of genetically-engineered glyphosate-resistant wheat was found on a farm in Oregon. Extensive testing confirmed the wheat as a variety – MON71800. The wheat had been developed by Monsanto but never been approved or marketed after the company had tested it between 1998 and 2005. The unexplained presence of this type of wheat presents a problem to wheat growers when buyers demand GMO-free wheat.

Japan subsequently suspended import of soft white wheat from the United States. A Kansas farmer sued Monsanto over the release, saying it had caused the price of wheat grown in the US to fall. Monsanto suggested that the presence of this wheat was likely an act of sabotage. On Jun 14, 2013, the USDA announced: "As of today, USDA has neither found nor been informed of anything that would indicate that this incident amounts to more than a single isolated incident in a single field on a single farm. All information collected so far shows no indication of the presence of GE wheat in commerce". As of August 30, 2013, while the source of the GM wheat remained unknown, Japan, South Korea and Taiwan had all resumed placing orders, and the disruption of the export market was minimal.

The investigation was closed in 2014 after the APHIS had exhausted all leads but had not found any evidence that the wheat had entered commercial supply.

In 2019, the USDA announced that genetically modified wheat plants engineered to resist Roundup were detected in an unplanted field in Washington state.

Genetically Modified Tomato

A genetically modified tomato, or transgenic tomato, is a tomato that has had its genes modified, using genetic engineering. The first trial genetically modified food was a tomato engineered to have a longer shelf life (the Flavr Savr), but never made it to market. Currently there are no genetically modified tomatoes available commercially, but scientists are developing tomatoes with new traits like increased resistance to pests or environmental stresses. Other projects aim to enrich tomatoes with substances that may offer health benefits or be more nutritious. As well as aiming to produce novel crops, scientists produce genetically modified tomatoes to understand the function of genes naturally present in tomatoes.

Plant physiologist Athanasios Theologis with tomatoes
that contain the bioengineered ACC synthase gene.

Wild tomatoes are small, green and largely unappetizing, but after centuries of breeding there are now thousands of varieties grown worldwide. Agrobacterium-mediated genetic engineering techniques were developed in the late 1980s that could successfully transfer genetic material into the nuclear genome of tomatoes. Genetic material can also be inserted into a tomato cell's chloroplast and chromoplast plastomes using biolistics. Tomatoes were the first food crop with an edible fruit where this was possible.

Delayed Ripening

Tomatoes have been used as a model organism to study the fruit ripening of climacteric fruit. To understand the mechanisms involved in the process of ripening, scientists have genetically engineered potatoes.

In 1994, the Flavr Savr became the first commercially grown genetically engineered food to be granted a license for human consumption. A second copy of the tomato gene *polygalacturonase* was inserted into the tomato genome in the antisense direction. The polygalacturonase enzyme degrades pectin, a component of the tomato cell wall, causing the fruit to soften. When the antisense gene is expressed it interferes with the production of the polygalacturonase enzyme, delaying the ripening process. The Flavr Savr failed to achieve commercial success and was withdrawn from the market in 1997. Similar technology, but using a truncated version of the polygalacturonase gene, was used to make a tomato paste.

DNA Plant Technology (DNAP), Agritope and Monsanto developed tomatoes that delayed ripening by preventing the production of ethylene, a hormone that triggers ripening of fruit. All three tomatoes inhibited ethylene production by reducing the amount of 1-aminocyclopropane-1-carboxylic acid (ACC), the precursor to ethylene. DNAP's tomato, called Endless Summer, inserted a truncated version of the *ACC synthase* gene into the tomato that interfered with the endogenous *ACC synthase*. Monsanto's tomato was engineered with the *ACC deaminase* gene from the soil bacterium *Pseudomonas chlororaphis* that lowered ethylene levels by breaking down ACC. Agritope introduced an S-adenosylmethionine hydrolase (SAMase) encoding gene derived from the *E. coli* bacteriophage T3, which reduced the levels of S-adenosylmethionine, a precursor to ACC. Endless Summer was briefly tested in the marketplace, but patent arguments forced its withdrawal.

Scientists in India have delayed the ripening of tomatoes by silencing two genes encoding N-glycoprotein modifying enzymes, α-mannosidase and β-D-N-acetylhexosaminidase. The fruits produced were not visibly damaged after being stored at room temperature for 45 days, whereas unmodified tomatoes had gone rotten. In India, where 30% of fruit is wasted before it reaches the market due to a lack of refrigeration and poor road infrastructure, the researchers hope genetic engineering of the tomato may decrease wastage.

Environmental Stress Tolerance

Abiotic stresses like frost, drought and increased salinity are a limiting factor to the growth of tomatoes. While no genetically modified stress-tolerant plants are currently commercialised, transgenic approaches have been researched. An early tomato was developed that contained an antifreeze gene (*afa3*) from the winter flounder with the aim of increasing the tomato's tolerance to frost, which became an icon in the early years of the debate over genetically modified foods, especially in relation to the perceived ethical dilemma of combining genes from different species. This tomato gained the moniker "fish tomato". The antifreeze protein was found to inhibit ice recrystallization in the flounder blood, but had no effect when expressed in transgenic tobacco. The resulting tomato was never commercialized, possibly because the transgenic plant did not perform well in its frost-tolerance or other agronomic characteristics.

Other genes from various species have been inserted into the tomato with the hope of increasing their resistance to various environmental factors. A gene from rice (*Osmyb4*), which codes for a transcription factor, that was shown to increase cold and drought tolerance in transgenic *Arabidopsis thaliana* plants, was inserted into the tomato. This resulted in increased drought tolerance, but did not appear to have any effect on cold tolerance. Overexpressing a vacuolar Na^+/H^+ antiport (*AtNHX1*) from *A. thaliana* lead to salt accumulating in the leaves of the plants, but not in the fruit and allowed them to grow more in salt solutions than wildtype plants. They were the first salt-tolerant, edible plants ever created. Tobacco osmotic genes overexpressed in tomatoes produced plants that held a higher water content than wildtype plants increasing tolerance to drought and salt stress.

Pest Resistance

The insecticidal toxin from the bacterium *Bacillus thuringiensis* has been inserted into a tomato plant. When field tested they showed resistance to the tobacco hornworm (*Manduca sexta*), tomato fruitworm (*Heliothis zea*), the tomato pinworm (*Keiferia lycopersicella*) and the tomato fruit borer (*Helicoverpa armigera*). A 91-day feeding trial in rats showed no adverse effects, but the Bt tomato has never been commercialised. Tomatoes resistant to a root knot nematode have been created by inserting a cysteine proteinase inhibitor gene from taro. A chemically synthesised *cecropin B* gene, usually found in the giant silk moth (*Hyalophora cecropia*), has been introduced into tomato plants and in vivo studies show significant resistance to bacterial wilt and bacterial spot. When the cell wall proteins, polygalacturonase and expansin are prevented from being produced in fruits, they are less susceptible to the fungus *Botrytis cinerea* than normal tomatoes. Pest resistant tomatoes can reduce the ecological footprint of tomato production while at the same time increase farm income.

Improved Nutrition

Tomatoes have been altered in attempts to add nutritional content. In 2000, the concentration of pro-vitamin A was increased by adding a bacterial gene encoding phytoene desaturase, although the total amount of carotenoids remained equal. The researchers admitted at the time that it had no prospect of being grown commercially due to the anti-GM climate. Sue Meyer of the pressure group Genewatch, told *The Independent* that she believed, "If you change the basic biochemistry, you could alter the levels of other nutrients very important for health". More recently, scientists created blue tomatoes that have increased the production of anthocyanin, an antioxidant in tomatoes in several ways. One group added a transcription factor for the production of anthocyanin from *Arabidopsis thaliana* whereas another used transcription factors from snapdragon (*Antirrhinum*). When the snapdragon genes were used, the fruits had similar anthocyanin concentrations to blackberries and blueberries. The inventors of the GMO blue tomato using snapdragon genes, Jonathan Jones and Cathie Martin of the John Innes Centre, founded a company called Norfolk Plant Sciences to commercialize the blue tomato. They partnered with a company in Canada called New Energy Farms to grow a large crop of blue tomatoes, from which to create juice to test in clinical trials on the way to obtaining regulatory approval.

Another group has tried to increase the levels of isoflavone, known for its potential cancer preventive properties, by introducing the soybean *isoflavone synthase* into tomatoes.

Improved Taste

When geraniol synthase from lemon basil (*Ocimum basilicum*) was expressed in tomato fruits under a fruit-specific promoter, 60% of untrained taste testers preferred the taste and smell of the transgenic tomatoes. The fruits contained around half the amount of lycopene, reducing the health benefits of eating them.

Vaccines

Tomatoes (along with potatoes, bananas and other plants) are being investigated as vehicles for delivering edible vaccines. Clinical trials have been conducted on mice using tomatoes expressing antibodies or proteins that stimulate antibody production targeted to norovirus, hepatitis B, rabies, HIV, anthrax and respiratory syncytial virus. Korean scientists are looking at using the tomato to express a vaccine against Alzheimer's disease. Hilary Koprowski, who was involved in the development of the polio vaccine, led a group of researchers in developing a tomato expressing a recombinant vaccine to SARS.

Genetically Modified Maize

Genetically modified maize (corn) is a genetically modified crop. Specific maize strains have been genetically engineered to express agriculturally-desirable traits, including resistance to pests and to herbicides. Maize strains with both traits are now in use in multiple countries. GM maize has also caused controversy with respect to possible health effects, impact on other insects and impact on other plants via gene flow. One strain, called Starlink, was approved only for animal feed in the US but was found in food, leading to a series of recalls starting in 2000.

Transgenic maize containing a gene from the bacteria *Bacillus thuringiensis.*

Herbicide Resistant Maize

Corn varieties resistant to glyphosate herbicides were first commercialized in 1996 by Monsanto, and are known as "Roundup Ready Corn". They tolerate the use of Roundup. Bayer CropScience developed "Liberty Link Corn" that is resistant to glufosinate. Pioneer Hi-Bred has developed and markets corn hybrids with tolerance to imidazoline herbicides under the trademark "Clearfield" – though in these hybrids, the herbicide-tolerance trait was bred using tissue culture selection and the chemical mutagen ethyl methanesulfonate, not genetic engineering. Consequently, the regulatory framework governing the approval of transgenic crops does not apply for Clearfield.

As of 2011, herbicide-resistant GM corn was grown in 14 countries. By 2012, 26 varieties of herbicide-resistant GM maize were authorised for import into the European Union., but such imports remain controversial. Cultivation of herbicide-resistant corn in the EU provides substantial farm-level benefits.

Insect-resistant Corn

The European corn borer, *Ostrinia nubilalis,* destroys corn crops by burrowing into the stem, causing the plant to fall over.

Bt corn is a variant of maize that has been genetically altered to express one or more proteins from the bacterium *Bacillus thuringiensis* including Delta endotoxins. The protein is poisonous to certain insect pests. Spores of the bacillus are widely used in organic gardening, although GM corn is not considered organic. The European corn borer causes about a billion dollars in damage to corn crops each year.

In recent years, traits have been added to ward off corn ear worms and root worms, the latter of which annually causes about a billion dollars in damages.

The Bt protein is expressed throughout the plant. When a vulnerable insect eats the Bt-containing plant, the protein is activated in its gut, which is alkaline. In the alkaline environment, the protein partially unfolds and is cut by other proteins, forming a toxin that paralyzes the insect's digestive system and forms holes in the gut wall. The insect stops eating within a few hours and eventually starves.

In 1996, the first GM maize producing a Bt Cry protein was approved, which killed the European corn borer and related species; subsequent Bt genes were introduced that killed corn rootworm larvae.

Approved Bt genes include single and stacked (event names bracketed) configurations of: Cry1A.105 (MON89034), CryIAb (MON810), CryIF (1507), Cry2Ab (MON89034), Cry3Bb1 (MON863 and MON88017), Cry34Ab1 (59122), Cry35Ab1 (59122), mCry3A (MIR604), and Vip3A (MIR162), in both corn and cotton. Corn genetically modified to produce VIP was first approved in the US in 2010.

In 2018 a study found that Bt-corn protected nearby fields of non-Bt corn and nearby vegetable crops, reducing the use of pesticides on those crops. Data from 1976-1996 (before Bt corn was widespread) was compared to data after it was adopted. They examined levels of the European corn borer and corn earworm. Their larvae eat a variety of crops, including peppers and green beans. Between 1992 and 2016, the amount of insecticide applied to New Jersey pepper fields decreased by 85 percent. Another factor was the introduction of more effective pesticides that were applied less often.

Sweet Corn

GM sweet corn varieties include "Attribute", the brand name for insect-resistant sweet corn developed by Syngenta and Performance Series insect-resistant sweet corn developed by Monsanto.

Drought Resistance Maize

In 2013 Monsanto launched the first transgenic drought tolerance trait in a line of corn hybrids called DroughtGard. The MON 87460 trait is provided by the insertion of the cspB gene from the soil microbe *Bacillus subtilis*; it was approved by the USDA in 2011 and by China in 2013.

Maize with Increased Nutritional Value

Research has been done on adding a single *E. coli* gene to maize to enable it to be grown with an essential amino acid (methionine).

Health Safety

In regular corn crops, insects promote fungal colonization by creating "wounds", or holes, in corn kernels. These wounds are favored by fungal spores for germination, which subsequently leads to mycotoxin accumulation in the crop that can be carcinogenic and toxic to humans and other animals. This can prove to be especially devastating in developing countries with drastic climate patterns such as high temperatures, which favor the development of toxic fungi. In addition, higher mycotoxin levels leads to market rejection or reduced market prices for the grain. GM corn crops encounter fewer insect attacks, and thus, have lower concentrations of mycotoxins. Fewer insect attacks also keep corn ears from being damaged, which increases overall yields.

Resistance

Resistant strains of the European corn borer have developed in areas with defective or absent refuge management.

In November 2009, Monsanto scientists found the pink bollworm had become resistant to first-generation Bt cotton in parts of Gujarat, India – that generation expresses one Bt gene, *Cry1Ac*. This was the first instance of Bt resistance confirmed by Monsanto anywhere in the world. Bollworm resistance to first generation Bt cotton has been identified in Australia, China, Spain, and the United States. In 2012, a Florida field trial demonstrated that army worms were resistant to pesticide-containing GM corn produced by Dupont-Dow; armyworm resistance was first discovered in Puerto Rico in 2006, prompting Dow and DuPont to voluntarily stop selling the product on the island.

Controversy

There is a scientific consensus that currently available food derived from GM crops poses no greater risk to human health than conventional food, but that each GM food needs to be tested on a case-by-case basis before introduction. Nonetheless, members of the public are much less likely than scientists to perceive GM foods as safe. The legal and regulatory status of GM foods varies by country, with some nations banning or restricting them, and others permitting them with widely differing degrees of regulation.

The scientific rigor of the studies regarding human health has been disputed due to alleged lack of independence and due to conflicts of interest involving governing bodies and some of those who perform and evaluate the studies.

GM crops provide a number of ecological benefits, but there are also concerns for their overuse, stalled research outside of the Bt seed industry, proper management and issues with Bt resistance arising from their misuse.

Critics have objected to GM crops on ecological, economic and health grounds. The economic issues derive from those organisms that are subject to intellectual property law, mostly patents. The first generation of GM crops lose patent protection beginning in 2015. Monsanto has claimed it will not pursue farmers who retain seeds of off-patent varieties. These controversies have led to litigation, international trade disputes, protests and to restrictive legislation in most countries.

Effects on Nontarget Insects

Critics claim that Bt proteins could target predatory and other beneficial or harmless insects as well as the targeted pest. These proteins have been used as organic sprays for insect control in France since 1938 and the USA since 1958 with no ill effects on the environment reported. While *cyt* proteins are toxic towards the insect order Diptera (flies), certain *cry* proteins selectively target lepidopterans (moths and butterflies), while other *cyt* selectively target Coleoptera. As a toxic mechanism, *cry* proteins bind to specific receptors on the membranes of mid-gut (epithelial) cells, resulting in rupture of those cells. Any organism that lacks the appropriate gut receptors cannot be affected by the *cry* protein, and therefore Bt. Regulatory agencies assess the potential for the transgenic plant to impact nontarget organisms before approving commercial release.

A 1999 study found that in a lab environment, pollen from Bt maize dusted onto milkweed could harm the monarch butterfly. Several groups later studied the phenomenon in both the field and the laboratory, resulting in a risk assessment that concluded that any risk posed by the corn to butterfly populations under real-world conditions was negligible. A 2002 review of the scientific literature concluded that "the commercial large-scale cultivation of current Bt–maize hybrids did not pose a significant risk to the monarch population". A 2007 review found that "nontarget invertebrates are generally more abundant in Bt cotton and Bt maize fields than in nontransgenic fields managed with insecticides. However, in comparison with insecticide-free control fields, certain nontarget taxa are less abundant in Bt fields".

Gene Flow

Gene flow is the transfer of genes and alleles from one species to another. Concerns focus on the interaction between GM and other maize varieties in Mexico, and of gene flow into refuges.

In 2009 the government of Mexico created a regulatory pathway for genetically modified maize, but because Mexico is the center of diversity for maize, gene flow could affect a large fraction of the world's maize strains. A 2001 report in *Nature* presented evidence that Bt maize was cross-breeding with unmodified maize in Mexico. The data in this paper was later described as originating from an artifact. *Nature* later stated, "the evidence available is not sufficient to justify the publication of the original paper". A 2005 large-scale study failed to find any evidence of contamination in Oaxaca. However, other Researchers also found evidence of cross-breeding between natural maize and transgenic maize.

A 2004 study found Bt protein in kernels of refuge corn.

In 2017, a large-scale study found "pervasive presence of transgenes and glyphosate in maize-derived food in Mexico".

References

- Craig Freudenrich; Dora Barlaz; Jane Gardner (2009). AP Environmental Science. Kaplen inc. pp. 189–190. ISBN 978-1-4277-9816-9
- Wilhelm, Steve (26 July 2012). "Non-browning GMO apple resisted by Washington state growers associations". Puget Sound Business Journal. Puget Sound Business Journal. Retrieved November 6, 2020
- Gunther, Marc (27 June 2007). "Genetically Engineered Rice Gets into the U.S. Food Supply". CNNMoney. Retrieved 11 November 2011
- Healthy-foods, Diet-and-nutrition, The-negative-health-effects-of-Genetically-Modified-foods-and-how-to-negate-them-20141022: health24.com, Retrieved 09 January, 2020
- Dawe, D.; Robertson, R.; Unnevehr, L. (2002). "Golden rice: what role could it play in alleviation of vitamin A deficiency?". Food Policy. 27 (5–6): 541–60. doi:10.1016/S0306-9192(02)00065-9

PERMISSIONS

All chapters in this book are published with permission under the Creative Commons Attribution Share Alike License or equivalent. Every chapter published in this book has been scrutinized by our experts. Their significance has been extensively debated. The topics covered herein carry significant information for a comprehensive understanding. They may even be implemented as practical applications or may be referred to as a beginning point for further studies.

We would like to thank the editorial team for lending their expertise to make the book truly unique. They have played a crucial role in the development of this book. Without their invaluable contributions this book wouldn't have been possible. They have made vital efforts to compile up to date information on the varied aspects of this subject to make this book a valuable addition to the collection of many professionals and students.

This book was conceptualized with the vision of imparting up-to-date and integrated information in this field. To ensure the same, a matchless editorial board was set up. Every individual on the board went through rigorous rounds of assessment to prove their worth. After which they invested a large part of their time researching and compiling the most relevant data for our readers.

The editorial board has been involved in producing this book since its inception. They have spent rigorous hours researching and exploring the diverse topics which have resulted in the successful publishing of this book. They have passed on their knowledge of decades through this book. To expedite this challenging task, the publisher supported the team at every step. A small team of assistant editors was also appointed to further simplify the editing procedure and attain best results for the readers.

Apart from the editorial board, the designing team has also invested a significant amount of their time in understanding the subject and creating the most relevant covers. They scrutinized every image to scout for the most suitable representation of the subject and create an appropriate cover for the book.

The publishing team has been an ardent support to the editorial, designing and production team. Their endless efforts to recruit the best for this project, has resulted in the accomplishment of this book. They are a veteran in the field of academics and their pool of knowledge is as vast as their experience in printing. Their expertise and guidance has proved useful at every step. Their uncompromising quality standards have made this book an exceptional effort. Their encouragement from time to time has been an inspiration for everyone.

The publisher and the editorial board hope that this book will prove to be a valuable piece of knowledge for students, practitioners and scholars across the globe.

INDEX

www.ingramcontent.com/pod-product-compliance
Lightning Source LLC
LaVergne TN
LVHW080622210926
848485LV00003B/198
9781639874316